Advanced Welding Techniques

To meet weight, quality, and cost targets, it is essential to design, develop, and manufacture optimal, cost-effective welded structures that take into consideration material, process, and dimensioning procedures. For effective design, a weld designer must have a comprehensive grasp of welding basics, associated metallurgy, and fabrication and characterization processes. *Advanced Welding Techniques* highlights breakthroughs in advances in welding methods and provides readers with the ability to accurately identify the appropriate welding processes and optimal improvement methods for intended applications. It offers comprehensive guidance on welding design to ensure readers are equipped to provide solutions to any technical malfunctions they may encounter, including:

- Supplies essential stepwise knowledge on design for welding, starting with the fundamentals to the complex
- Covers the role of filler metals and parameters on welding performance, emerging and advanced welding techniques, and advantages and limitations of various methods
- Discusses integration of additive manufacturing and welding
- Contains practical applications
- Considers challenges and future scope for further research as well as future challenges

This book offers students, academics, researchers, scientists, engineers, and industry experts a comprehensive overview of the most recent breakthroughs in advanced welding methods and their applications to joining various metals and their alloys.

Himanshu Vashishtha is postdoctoral research associate at the University of Cambridge, UK, working on additively manufactured cardiovascular stents and synchrotron beamlines. He was previously a research fellow at the University of Birmingham, UK and the Indian Institute of Delhi, India. He completed his master's and doctoral degrees from Visvesvaraya National Institute of Technology in Nagpur, India.

Deepak Kumar is postdoctoral research associate in the Department of Mechanical Engineering at Carnegie Mellon University, USA, where he is developing conductive oxides as contact materials for micro- and nanoelectromechanical switches. He received his doctorate from the Materials Science and Engineering Department of the Indian Institute of Technology in Delhi, India.

Ravindra Vasantro Taiwade is associate professor in the Metallurgical and Materials Engineering Department at Visvesvaraya National Institute of Technology Nagpur, India. He also received his Ph.D. from Visvesvaraya National Institute of Technology (VNIT) in Nagpur.

Advanced Welding Techniques

Current Trends and Future Perspectives

Edited by
Himanshu Vashishtha, Deepak Kumar and
Ravindra V. Taiwade

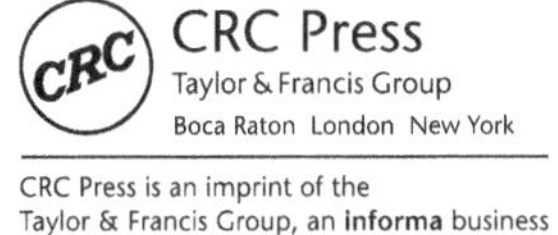

CRC Press is an imprint of the
Taylor & Francis Group, an informa business

Designed cover image: © Shutterstock, Gorodenkoff

First edition published 2025
by CRC Press
2385 NW Executive Center Drive, Suite 320, Boca Raton FL 33431

and by CRC Press
4 Park Square, Milton Park, Abingdon, Oxon, OX14 4RN

CRC Press is an imprint of Taylor & Francis Group, LLC

Library of Congress Cataloging-in-Publication Data
Names: Vashishtha, Himanshu author, editor.
Title: Advanced welding techniques : current trends and future perspectives /
edited by Himanshu Vashishtha, Deepak Kumar, Ravindra V. Taiwade.
Description: First edition. | Boca Raton, FL: CRC Press, 2025. |
Includes bibliographical references and index. |
Summary: "To meet weight, quality, and cost targets, it is essential to design, develop, and manufacture optimal, cost-effective welded structures that take into consideration material, process, and dimensioning procedures. For effective design, a weld designer must have a comprehensive grasp of welding basics, associated metallurgy, and fabrication and characterization processes. Advanced Welding Techniques highlights breakthroughs in advances in welding methods and provides readers with the ability to accurately identify the appropriate welding processes and optimal improvement methods for intended applications. It offers comprehensive guidance on welding design to ensure readers are equipped to provide solutions to any technical malfunctions they may encounter. Supplies essential stepwise knowledge on design for welding, starting with the fundamentals to the complex. Covers role of filler metals and parameters on welding performance, emerging and advanced welding techniques, and advantages and limitations of various methods. Discusses integration of additive manufacturing and welding. Contains practical applications. Considers challenges and future scope for further research as well as future challenges. This book offers students, academics, researchers, scientists, engineers, and industry experts a comprehensive overview the most recent breakthroughs in advanced welding methods and their applications to joining various metals and their alloys"–Provided by publisher.
Identifiers: LCCN 2024016628 (print) | LCCN 2024016629 (ebook) |
ISBN 9781032565118 (hardback) | ISBN 9781032565132 (paperback) | ISBN 9781003435884 (ebook)
Subjects: LCSH: Welding–Technological innovations.
Classification: LCC TS227.2 .A375 2024 (print) | LCC TS227.2 (ebook) |
DDC 671.5/2–dc23/eng/20240610
LC record available at https://lccn.loc.gov/2024016628
LC ebook record available at https://lccn.loc.gov/2024016629

ISBN: 978-1-032-56511-8 (hbk)
ISBN: 978-1-032-56513-2 (pbk)
ISBN: 978-1-003-43588-4 (ebk)

DOI: 10.1201/9781003435884

Typeset in Times
by Newgen Publishing UK

Contents

Preface vii
Contributors ix

Chapter 1 Welding Scope, Application, and Opportunities 1

Jagesvar Verma and Ravindra V. Taiwade

Chapter 2 Various Advanced Welding Methods: A Brief Overview 22

Gaurav

Chapter 3 Heat Generation and Physics of the Arc Welding Process 38

Pavan Meena, Ramkishor Anant, and Nilesh Kumar Paraye

Chapter 4 Defects Associated with Welding Techniques and Their Detection Methods 63

Ranjeet Kumar

Chapter 5 Role of Filler Materials on Welding Performance 90

Nilesh Kumar Paraye, Ramkishor Anant, and Himanshu Vashishtha

Chapter 6 Effect of Welding Parameters on Structure-Property Correlation 115

Santosh K. Gupta, Vipin Tandon, and Himanshu Vashishtha

Chapter 7 Tungsten and Metal Inert Gas Welding: Automation in Welding 136

Virendra Pratap Singh, Basil Kuriachen, Vinyas Mahesh, and Dineshkumar Harursampath

Chapter 8 Friction Stir Welding and Design 153

Namdev Ashok Patil, Roshan Vijay Marode, and Sambhaji Kashinath Kusekar

Chapter 9 Laser Beam Welding: Opportunity and Challenges 167

Atul Kumar, Anoj Giri, Ashish Kumar Saxena, and Rishav Kumar

Chapter 10 Hybrid Welding Processes: Challenges and Future Perspective 180

Brij Mohan Mundotiya

Chapter 11 Additive Manufacturing Integration with Welding 198

Laukik P. Raut, Ravindra V. Taiwade, Ashish Fande, Dhiraj Narayane, and Prachi Tawele

Chapter 12 Current Scenario, Future Scope, and Challenges in Welding........... 211

Nitin Kumar Lautre

Index 233

Preface

In the realm of industrial fabrication and construction, welding stands as a cornerstone, holding together the intricate structures that define our modern world. Welding overcomes the issues associated with the complicated production and transporting of large systems. Since the Industrial Revolution, several welding techniques and equipment have been created. Specifically, in the last four decades, welding technology has seen extensive technical changes and advancements. The emergence of diverse materials to meet the industrial needs of the twenty-first century is attributable to the advancement of complicated welding methods. The goal of the editors is to provide the readers with the ability to accurately identify the appropriate welding process for their applications and to select the optimal improvement method. Choosing the appropriate weld method for the applications depends on the engineering, design, process, and customer requirements for the best weld to utilize on each joint. Therefore, there is a need to develop an understanding of the various welding processes from basic to advanced levels, their limitations and advantages, which in turn leads to further advancement in the area of welding. This book also serves as a comprehensive exploration into the evolving landscape of welding, delving into both established methodologies and cutting-edge advancements that promise to reshape the industry. Welding, once perceived as a straightforward joining process, has undergone a remarkable transformation in recent years. Technological innovations, coupled with a deeper understanding of metallurgy and material sciences, have propelled welding into a realm of unparalleled sophistication. As industries demand higher efficiency, durability, and precision, welders are challenged to push the boundaries of traditional techniques and embrace new methodologies.

Advanced Welding Techniques is a culmination of expertise from seasoned professionals, researchers, and innovators at the forefront of welding technology. Through a meticulous examination of current trends, we navigate through the complexities of processes such as laser welding, friction stir welding, and electron beam welding, among others. We uncover the intricacies of metallurgical considerations, non-destructive testing methods, and automation solutions that are reshaping the landscape of welding practices. Furthermore, this book ventures into uncharted territories, presenting forward-looking insights into the future of welding. This book explores how emerging technologies are poised to revolutionize the way we weld and fabricate structures. Whether you are a seasoned welding professional seeking to expand your knowledge or an aspiring enthusiast eager to embark on a journey into the world of metal joining, *Advanced Welding Techniques* offers a comprehensive roadmap, and serves as a fulcrum for all those concerned in welding science, providing them with adequate information to explore this fascinating landscape. It is our fervent hope that this book aids as a beacon of inspiration, guiding readers toward mastery in an ever-evolving craft.

Contributors

Ramkishor Anant
Department of MME
Maulana Azad National Institute of Technology
Bhopal (MP), India

Ashish Fande
Department of Mechanical Engineering
Nagpur Institute of Technology
Nagpur, India

Gaurav
Department of Mechanical Engineering
Government Engineering College
Buxar, India

Department of Applied Mechanics
Indian Institute of Technology Delhi
New Delhi, India

Anoj Giri
School of Mechanical Engineering
Vellore Institute of Technology
Vellore, India

Santosh K. Gupta
Department of Metallurgical and Materials Engineering
Visvesvaraya National Institute of Technology
Nagpur, India

Dineshkumar Harursampath
NMCAD Lab.
Department of Aerospace Engineering
Indian Institute of Science
Bengaluru, India

Atul Kumar
School of Mechanical Engineering
Vellore Institute of Technology
Vellore, India

Ranjeet Kumar
Materials Engineering Division CSIR-National Metallurgical Laboratory
Jamshedpur, India

Rishav Kumar
School of Mechanical Engineering
Vellore Institute of Technology
Vellore, India

Sambhaji Kashinath Kusekar
Department of Mechanical Engineering
Cleveland State University
Cleveland, Ohio, USA

Basil Kuriachen
Department of Mechanical Engineering
National Institute of Technology Calicut
Kerala, India

Nitin Kumar Lautre
Department of Mechanical Engineering
Visvesvaraya National Institute of Technology (VNIT)
Nagpur, Maharashtra, India

Vinyas Mahesh
Department of Mechanical Engineering
National Institute of Technology Silchar
Assam, India

Roshan Vijay Marode
Department of Mechanical Engineering
Universiti Teknologi PETRONAS
Malaysia

Pavan Meena
Department of MME
Maulana Azad National Institute of Technology
Bhopal (MP) Republic of India

Brij Mohan Mundotiya
Department of Metallurgical and Materials Engineering
Malaviya National Institute of Technology Jaipur
Jaipur, Rajasthan, India

Dhiraj Narayane
Department of MME
VNIT
Nagpur, India

Nilesh Kumar Paraye
Institut De Recherche Dupuy De Lôme
Université Bretagne Sud
Lorient, France

Namdev Ashok Patil
Department of Product Design
Dr. D. Y. Patil Vidyapeeth
Dr. D. Y. Patil School of Design
Pune, India

Laukik P. Raut
Department of Mechanical Engineering
G. H. Raisoni College of Engineering
Nagpur, India

Ashish Kumar Saxena
Centre for Innovative Manufacturing Research
Vellore Institute of Technology
Vellore, India

Virendra Pratap Singh
Department of Mechanical Engineering
National Institute of Technology
Mizoram, India

Ravindra V. Taiwade
Department of Metallurgical and Materials Engineering
VNIT
Maharashtra, India

Vipin Tandon
Centre of Sustainable Built Environment
Manipal School of Architecture and Planning
Manipal Academy of Higher Education
Manipal, India

Prachi Tawele
Department of Mechanical Engineering
Priyadarshini College of Engineering
Nagpur, India

Himanshu Vashishtha
Department of Materials Science & Metallurgy
University of Cambridge
Cambridge, United Kingdom

Jagesvar Verma
Department of Mechanical and Manufacturing Engineering
National Institute of Advanced Manufacturing Technology
Ranchi, India

1 Welding Scope, Application, and Opportunities

Jagesvar Verma and Ravindra V. Taiwade

1. INTRODUCTION

Welding is a manufacturing process defined as the joining of two similar or dissimilar metals by heating and/or mixing at a temperature high enough to cause softening or melting, with or without the application of heat, pressure, and fillers [1]. Welding solves many complex problems of manufacturing and heavy assemblies and signifies an essential process in many industries from small products to very large projects [2]. In the last few years, the welding industry has grown very fast. In 2019 the global welding market size was US$ 19.53 billion, and it is anticipated to reach and surpass US$ 27.22 billion in 2027 [3,4]. The shipments of welding equipment are presumed to more than double in the next five years because the emphasis on welding as a prime basic manufacturing technology is increasing, hence the growth rate anticipated in the future will be approximately 8% per year [3,4]. The impactful industrial revolution and continued evolving new materials and technology around welding creates a lot of scope in the fields of welding and joining [5]. A large variety of welding processes, equipment, continued progress, and advancements are happening day to day [5, 6]. Innovation, advances, and progress in arc and other welding and joining processes continue to rise. Artificial intelligence- (AI) based welding techniques and robotized welding techniques were developed which completely changed the thoughts and working culture in manufacturing industries.

In the fusion welding process, tungsten inert gas welding (TIG) or gas tungsten arc welding (GTAW) is extensively used to join thin sections of stainless steel, but in case of thick sections in a single pass a lot of difficulties arise to maintain the sustainability in the TIG/GTAW process [7]. To maintain the sustainability, activated-TIG (A-TIG) was developed which improves the weld penetration 1.5–4 times in a single pass [7]; a thin activated flux layer which is a mixture of oxide or a halide powder and acetone or methanol applied to the plates improves weld penetration [7-9]. During welding, the paste evaporates, leaving a layer of the activated flux on the plate's surface [8,9]. There will be slots of energy savings in single-pass

DOI: 10.1201/9781003435884-1

A-TIG welding as compared to multi-pass TIG welding, however, it suffers from a high amount of slug due to entrapment of oxide flux, which can be minimized by two modifications such as flux bounded TIG (FB-TIG) and flux zone TIG (FZ-TIG) welding processes [7,10,11].

Similarly, advancements were made in friction stir welding (FSW) to weld thermoplastic materials, which was initially difficult to weld by any other welding process [12]. Friction stir welding was primarily developed for welding aluminium alloys then this method was successfully applied with several modifications to weld thermoplastic materials [13]. Recently, numerous external energy-assisted FSW techniques were developed [14] that are based on preheating the workpiece by an external energy source which overcomes the problems of conventional FSW processes for welding hard and high melting point materials. External energy is provided in the form of induction heat, laser heat, resistance heat, arc heat, radiation, and ultrasonic vibrations [14,15]. Significant advantages have been observed while using energy-assisted FSW such as extended process window, improved process parameters and mechanical properties, reduced load, and tool wear [14,16]. In the case of friction welding (FW), when joining a tube-to-tube plate, less power is consumed as compared with conventional TIG welding which reduces welding time, power consumption and provides a strong weld joint [17,18,19].

As modern manufacturing technology is growing with innovation and advancement, it becomes a foreseeable trend to realize automatic, robotic, flexible, and intelligent welding techniques [20]. The expansion of robotized welding technology is truly remarkable and is today one of the major application areas for industrial robots [21].

AI-based welding techniques were also developed and have been gaining popularity [22,23]. It digitized welding operations, welded part conditions, and welding equipment conditions to standardize and quantify man, material, and machine and to understand production conditions on a time axis and improve productivity in a shorter time [24]. Another advancement to date in the solid-state welding process is as high-velocity impact welding which is useful for dissimilar material welding that avoids the formation of intermetallic [25]. Five various techniques have been developed to date, namely gas gun welding (GGW), magnetic pulse welding (MPW), explosive welding (EXW), vaporizing foil actuator welding (VFAW), and laser impact welding (LIW) and all have a similar mechanism with different energy sources [26]. High-energy density beam welding is utilized in a large number of industries spanning medical products to aerospace and defence due to the high depth of penetration at travel speeds greater than other fusion welding processes [27].

In laser technology also many changes happened over the past 30 years. For many years a 10-micron wavelength CO_2 laser was king because it was versatile in maintaining high-quality welds. But over 10 years ago, high brightness 1-micron wavelength fiber and disk lasers came to the market with great promise, showing better and greater flexibility in laser processing, with an increase in energy efficiency [28]. A concentrated beam of electrons was first used for welding in a vacuum chamber in the 1950s in Germany and France, and in the past 50 years numerous applications of electron beams in surface material processing all over the world has been demonstrated [29].

This review effort mainly highlights the conventional and advanced welding processes based on heat source and pressure, recent work and development/innovation with scope, opportunities, and applications.

2. CLASSIFICATION OF WELDING TECHNOLOGY

Welding methods are classified based on heat source and applying pressure [30]. For welding, the source of energy is produced by applying energy in various forms to produce metallic bonds in pieces of metal using heat and/or applying pressure either at room temperature or at 0.50 to 0.90 times the rate of melting temperature [31]. Different welding processes based on the heat source and pressure [30,31] are illustrated in Figure 1.1 as Electrical energy (such as arc welding and resistance welding), Chemical energy (such as gas welding, explosive welding, thermite welding), Mechanical energy (such as friction stir welding, friction welding, and ultrasonic welding), and Radiation energy (like laser beam welding and electron beam welding) based processes.

2.1. Welding process based on heat source

Major welding processes based on heat sources are arc welding, resistance welding, gas welding, explosive welding, thermite welding, laser beam welding, and electron beam welding. However, several subcategories of mentioned welding processes are also used in various manufacturing sectors. This review focuses on the major welding processes.

2.1.1. Shielded arc welding

Shielded metal arc welding (SMAW) is the simplest, cheapest, and most widely used process as illustrated in Figure 1.2. It is also known as stick or manual metal arc welding (MMAW), especially in the United Kingdom and some European countries [32,33]. In this process, coalescence of metals is produced by heat from an electric

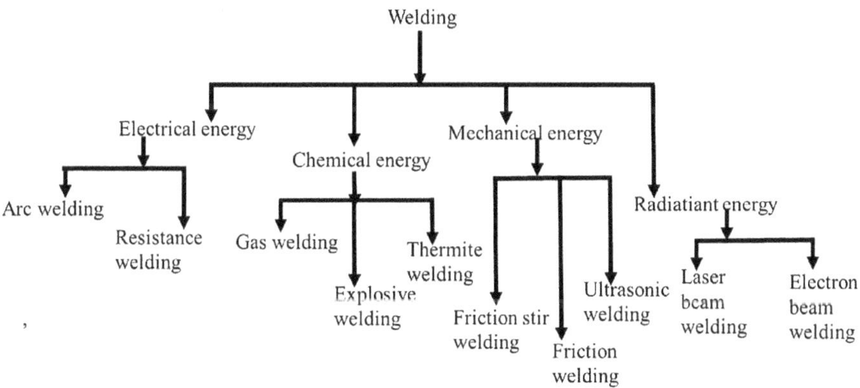

FIGURE 1.1 Classification of welding

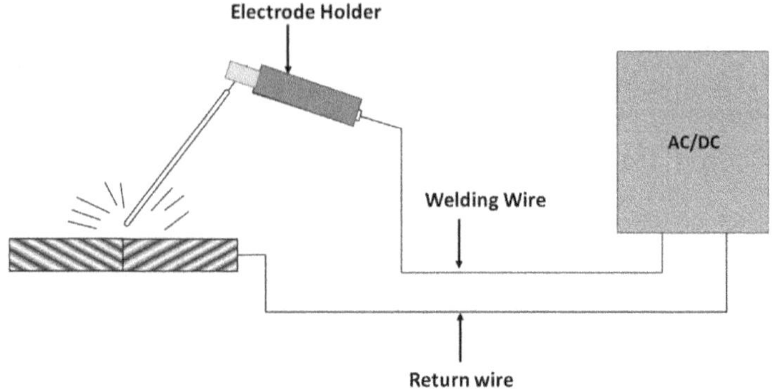

FIGURE 1.2 Schematic diagram of SMAW process

arc maintained between the tip of a consumable electrode and the base materials, consumable electrodes are mainly flux-covered [34]. The coating of electrodes primarily depends on a certain alloy and is mainly classified as slightly basic to slightly acidic [35]. The electrode may absorb moisture; to avoid the moisture it should be stored in a dry oven and maintained at approx. 120–200°C [35,6]. All electrodes in SMAW are suitable for alternating current (AC) or direct current (DC) mode but the recommended mode is direct current electrode positive (DCEP). A welding voltage of about 15–45 V and a welding current in the range of 10–500A are generally utilized to produce an arc with a temperature of 5000°C [34].

2.1.2. Gas tungsten arc welding

Gas tungsten arc welding is referred to as a tungsten inert gas welding and was developed in the late 1930s [37]. It is quite popular and the most widely used process due to good weld quality and cleaner welds; it is basically used for assembly and repairing the components (especially blade repair), applying the overlay, welding by root joints, and fill pass welding in several industries [38], but most application of GTAW is in the field of aeronautical gas turbines [39]. Gas tungsten arc welding is a manual process where a non-consumable tungsten electrode is used, and filler may be used separately along with a mixture of gases for shielding purposes (gas may be argon, carbon dioxide, helium, or the mixture of these gases). During welding, weld pool temperatures can reach 2500°C [40,41]. Figure 1.3a illustrates the GTAW process. The gas tungsten arc welding process is not limited only to the manual process, but it has several variants that involve different types of automation, which improve weld deposition rate [33,42] by a stabilized arc and good weld quality, mainly used for a thin workpiece up to 7 mm wall thickness. It requires a highly skilled operator and must be done at low speeds [33,42,43]. This process is suitable for welding stainless steel, aluminium, and other light materials [44]. Flux-assisted welding is recently developed as a variant of GTAW in which an activating flux is used to modify the weld pool to achieve a higher depth-to-width ratio; in a similar fashion, flux-zoned TIG, advanced activated-TIG welding, keyhole TIG welding, and Super-TIG welding are also developed [45].

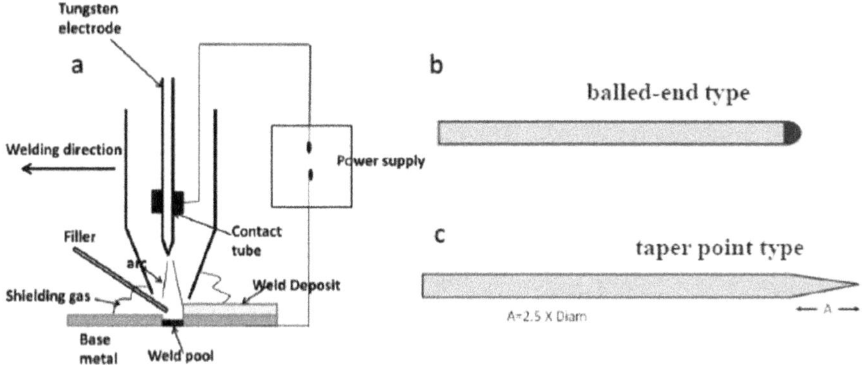

FIGURE 1.3 Schematic diagram of GTAW process

For better weld penetration a proper electrode tip preparation is essential, therefore, various shapes and coated non-consumable tungsten electrodes are being developed and used [46]. For nonferrous metals such as aluminium and magnesium welding, pure tungsten is used with a balled-end preparation on AC as shown in Figure 1.3(b) [47]. Whereas thoriated tungsten (1 or 2 percent thorium) is the most common electrode for carbon and stainless steels, it maintains the arc. It has a greater resistance to contamination and will maintain a sharp point with taper (Figure 1.3c) and will not break down as pure tungsten. Thoriated tungsten electrodes are ground to a point to give a better arc starting [46-48]. The degree of taper affects the shape and amount of penetration of a weld and the recommended taper length is two-and-a-half to three times the diameter of the electrode [46,48].

On the other hand, Zirconiated tungsten is mainly used on nonferrous metals for welding with higher AC currents [46,49].

2.1.3. Gas metal arc welding

Gas metal arc welding (GMAW) is a common name also referred to as metal inert gas welding (MIG) in which an arc is basically formed between a continuous, automatically fed, consumable electrode and workpiece in an inert atmosphere [50]. When inert gas is replaced by carbon dioxide then it is known as CO_2 arc welding or metal active gas (MAG) welding [50]. The GMAW process provides higher productivity and good weld quality, which lead to development in the arc welding process [51] and it has the potential to improve productivity over that obtained with the GTAW and SMAW processes due to its high operating factor and deposition rate [52]. The GMAW process can be done in all welding positions [53].

Gas metal arc welding is generally used in direct current electrode positive polarity (DCEP), in which the filler wire is directly connected to the positive terminal of the power source and the power source runs in the constant voltage (CV) mode [54]. Whereas reverse polarity gives stable arc, uniform metal transfer, and higher penetration [55]. Constant voltage power source can adjust the welding current in a way that the filler melting rate is equal to the given wire feed speed, and the welding voltage or

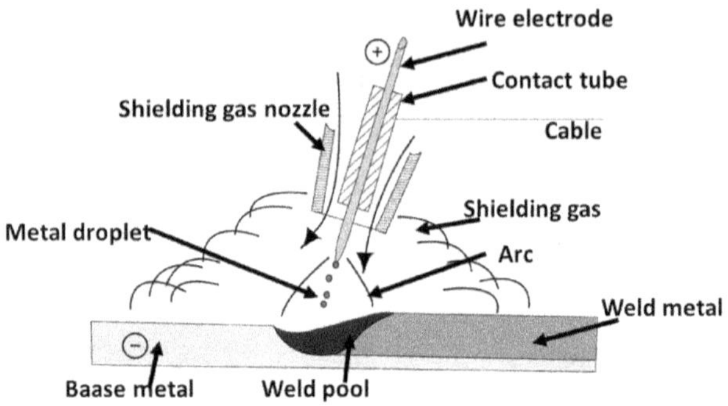

FIGURE 1.4 Schematic diagram of GMAW process

arc length is maintained constant [54, 55]. For automatic and semiautomatic welding, the productivity is frequently determined by the travel speed [54-56]. Faster travel speed requires a larger wire melting rate such that the melted metal is enough to form a longer weld bead in a unit of time [54,57].

Gas metal arc welding miserably fails when it is applied on thermo-susceptive metals like sheet-metals, sheathed thin plates. Burn through, porosity, excessive penetration, inadequate fusion, warping, and weld pool agitation, are the utmost typical challenges allied with the traditional GMAW processes. Figure 1.4 shows a schematic diagram of GMAW process with various nomenclature [58].

2.1.4. Resistance welding

Resistance welding is typically referred to as resistance spot welding (RSW), which is the most widely used and popular method in automotive and aerospace industries; it is applied in automobiles and the coaches of trains and planes, which reduces the cost and weight of the welded parts [60,61]. There are approx. 3000–6000 spot welds in any auto parts, which shows the significance of the resistance spot welding in particular fields [62]. Process gives better weld qualities, mechanical properties, structural performance, structural durability, safety design, stiffness, and strength [63]. In this process a very high rate of productivity can be achieved, for example welding 1+1 mm sheet, the process takes approx. 0.20 s only [64]. It is useful for welding thin sheets of Mg, Al, Ti, steel, and their alloys [63].

It is an inexpensive and more efficient welding technique [60]. It is based on a thermoelectric process (where heat and pressure are both applied) in which current flows through the workpieces held together and then pressure is applied by electrodes, and the heat generated by the resistance creates coalescence within a small area called nugget. The generation of heat mainly depends on the current between the workpieces, duration, and the resistance [65,66]. Energy conservation (Joule's law of heating) plays a role for heat generation from electric to thermal. The formula for heat generation is [61,63]:

$$\mathbf{H = I^2RT}$$

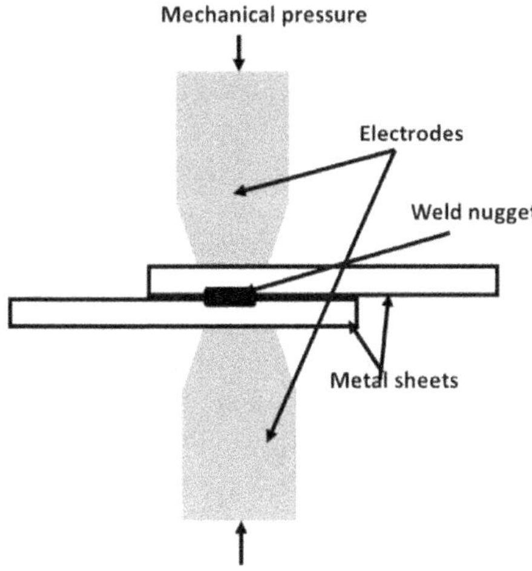

FIGURE 1.5 Working of spot welding process

Where H represents heat generation, I stands for an electric current, R is an electric resistance, and T is the time of current flow [61,63]. Figure 1.5 illustrated the resistance spot welding process. In RSW, the current is in the range of 1000–100000 ampere and the voltage is in the range 1–30 Volts to achieve joining [67]. To attain the desired current density, the proper electrode shape plays an important role; for this purpose, three main types of electrodes are commercially available such as flat type electrode, dome nose, and pointed types of electrodes [68]. The tip size of the electrode is deemed nearly equal to the nugget size, which is established by using Unwin's formula [68].

$$D_e = 6\sqrt{t}$$

Where t is the sheet thickness in mm and D_e is the electrode tip diameter. There are several different popular versions of resistance welding such as seam welding, percussion welding, flash butt welding, and projection welding [64].

2.1.5. Oxy-acetylene

Oxy-acetylene gas welding is popularly used for welding, brazing, soldering, and cutting operations for various metals and alloys with the help of the combustion of fuel gases such as acetylene, hydrogen, propane, or butane with the mixture of oxygen [69].

It is relatively inexpensive, portable, and versatile. The purpose of adopting this process by various industries is to produce a better quality product at minimum cost [69,70].

In this process the mixture of oxygen and acetylene gas provides a high temperature for welding. This process uses a shielding gas like argon to protect the weld pool from the outer contamination [71]. Filler is required to deposit excess metal to weld two parts of the base metals [50]. The acetylene gas and oxygen should be mixed in the correct required proportions and the welding torch can be adjusted in a way to produce the required type of flame to generate heat [72]. In this process, the temperature of approximately 3150°C is generated and heat is controlled by the pressure of the two gases by varying the size of the tip [73]. The cutting of metals and alloys requires more heat than that for welding, and for this purpose, oxidizing flame is utilized [74]. Three kinds of oxy-acetylene flame utilized based on the application are oxidizing flame, carburizing flame, and neutral flame [69]. Oxidizing flame is the flame in which the volume percentage of oxygen is more than that of acetylene in the torch; this oxidizing flame is mainly used for cutting metals/alloys. Whereas in carburizing flame the volume of acetylene is more than oxygen and mainly used for the pre- and post-heating purpose of welding. On the other hand, a neutral flame has a balanced proportion of oxygen and acetylene, used for welding metal and alloys [69].

2.1.6. Explosive welding

Explosive welding is a kind of hot-pressing technique and is commercially industrialized all over the world [75]. It was more popular and intensively used in the 1970s as an economically effective fabrication route especially for joining metal matrix and dissimilar metallurgical properties of metals which cannot be joined by any other conventional manufacturing process [76,77]. Nowadays it is also being used for cladding purposes [78]. Many researchers pointed out its excellent potential for the manufacture of claddings, in several sectors, including defence, aerospace, oil, gas, and chemical [79].

In this process the metal parts which are joined are arranged toward each other at an angle of 1–15°, which depends upon the material and method, and are prepared with a layer of explosive on the top. Detonation is used as an explosive material; after ignition, the top plate hits the bottom one at high speed approx. 100–1000 ms^{-1} against each other and joining happens at the contact area by local plastic deformation as illustrated in Figure 1.6 with various stages [78,80]. The cladding thickness may

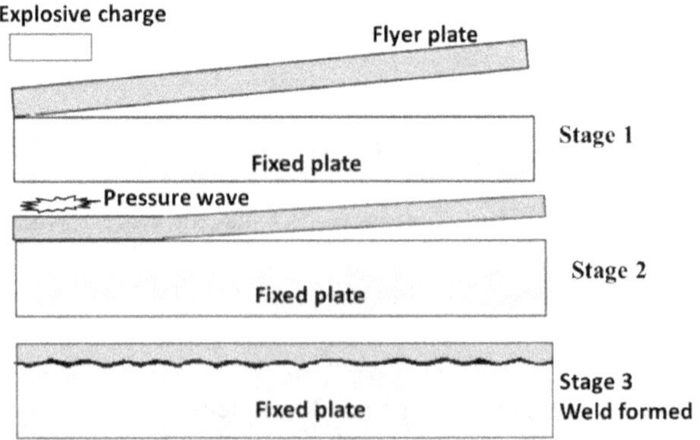

FIGURE 1.6 Basic steps of explosion welding

vary between 0.1 and 30 mm. During cladding detonation velocity reaches between 1200 and 7000 ms^{-1} and pressures are suitable in the range of 10 to 100 kbar [78].

2.1.7. Thermite welding

Thermite welding is a welding process which uses heat from exothermic reaction to create coalescence between metal pieces. The name is derived from "thermite," which is the generic name given to a reaction between metal oxides and reducing agents [33,81]. Thermite welding is more popular for welding rails to provide a smooth surface satisfactorily at a low cost, mainly used in track on-site to join rail steel and repair continuous welded rail [82]. Thermite welding equipment is very simple: during the exothermic reaction, aluminium reduces the iron oxide because aluminium has higher oxygen affinity and stronger molecular bonds with oxygen than iron [83]. The ratio of thermite powder mixtures usually is about three parts by weight of iron oxide to one part of aluminium, and reaction will take place according to the following equation:

Metal oxide + aluminium aluminium oxide + metal + heat [33,81]. Theoretically, the temperature generated from the reaction is about 3100°C [84].

2.1.8. Laser beam welding

Laser beam welding (LBW) was first industrialized in the 1970s [82,85], used in industries for several applications such as welding, cutting, cleaning metallic surfaces, altering the surface roughness, and measuring dimensions with high precision. Laser beam welding is widely used in aeronautical, shipbuilding, and automobile sectors which require high precision work [86]. Lasers are extremely accurate and precise, and advanced laser beams can be focused down to 1/1000 of a millimeter [86,87]. Laser beam welding has several advantages: it helps to increase production rates, reduce production costs, and provides accuracy and repeatability with high-quality weld (with micron-level precision) [88].

Welding was performed and a joint was produced between two workpieces [89], with the help of a concentrated heat source and a focused beam in the range of 10 µm–100 mm in width [90]. Laser machines operate at very high voltage power supply, this high voltage starts the flash lamps, resulting in emitting the light photons [90,91], which was are then absorbed by the atoms of ruby crystal and then electrons become excited to their higher energy level. When they return to their lower energy state again they emit a photon of light that further stimulates the atom of excited electrons and produces two photons. This process remains continues which provides a concentrated laser beam [90,91].

Laser welding is a high-power-density (1 MW/cm^2) fusion welding process that produces high aspect ratio welds with a relatively low heat input compared with any other arc welding process, resulting in small heat-affected zones. The weld spot size can vary between 0.2 and 13 mm. It's capable of welding a wide variety of materials, particularly carbon steels, HSLA steels, aluminium, titanium, and stainless steels [88].

The process has two different modes of operation: conduction mode and keyhole mode [90, 92]. The basic difference between them is a power density which is applied to the weld zone. If laser intensity is not adequate to cause boiling then the conduction takes place. Whereas in keyhole mode, beam intensity is a so high which causes vaporization and forms a keyhole in the molten pool [90,92]. In LBW no electrodes and/or fillers metals are necessary for welding, hence the weld is not prone to contamination and deformation [93]. Various laser sources are available for industrial welding applications such as Nd:YAG (Nd- Yttrium Aluminium Garnet), Nd:YV04(Yttrium Vanadate), Yb: Fiber (Ytterbium) and CO_2 laser [86].

2.1.9. Electron beam welding

Electron beam welding (EBW) has existed for over 60 years, but in the last decade demand has been rising in the various sectors in the industry [94]. It has a very high-power density of about 10^8 W/cm^2 at the focus of the beam and energy transfers by conduction of heat through the surface of the workpiece [95]. An electron beam released from a tungsten cathode (most commonly directly heated cathode) is accelerated by a very high voltage in the range of 10–200 kV and speed over 100 km/s that bombards the welded area with a focused electron beam, and having a diameter in the range of 0.1–0.8 mm causes the conversion of the kinetic energy of electrons to heat and form a strong weld joint [94,96]. High welding speed results in narrow welds and heat-affected zones (HAZ), therefore, workpiece distortion is very less.

The process is also capable of welding workpieces, with the keyhole technique, it can weld metal thickness from 0.01 mm to 250 mm of steel and up to 500 mm of aluminium and alloys [90]. It is a versatile technology. For example, different grades of steel can be joined to each other by means of electron beam welding as well as hardly weldable by other welding methods like refractory metals (tungsten, molybdenum, niobium) and chemically active metals (titanium, zirconium, beryllium) [90-94]. Electron beam welding is also able to join dissimilar metals, from the practical point of view a good example is electron beam welding of steel and aluminium which cannot be joined in the fusion welding process because of the formation of intermetallic. Figure 1.7 shows the various parts of electron beam welding [94,96].

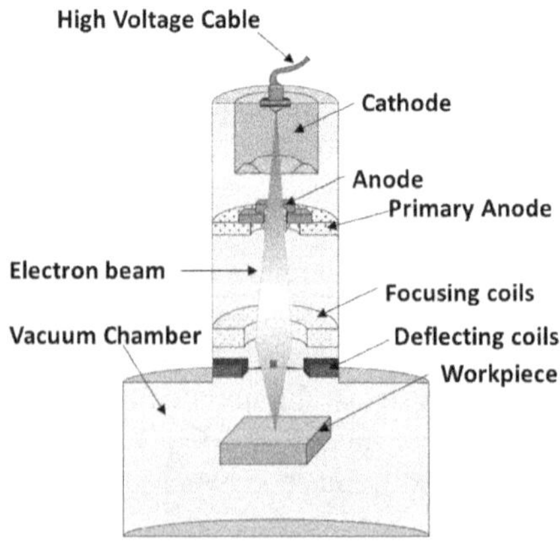

FIGURE 1.7 Sectional view of electron beam welding

2.2. Based on pressure applied

2.2.1. Friction stir welding

Friction stir welding is considered to be the most significant development in the solid-state metal joining process in a decade, which is energy-efficient and environment-friendly [97]. This process is versatile and particularly applies for joining high-strength aerospace grade aluminium and metallic alloys that are difficult to weld by conventional fusion welding processes in many industrial sectors such as marine, aerospace, railway and land transportation, and so on [98].

In the FSW process a non-consumable rotating tool is used which is designed in such a way that the pin and shoulder of the tool are introduced into the connecting edges of workpieces to be joined and travel along the weld path [99]. As the tool travels, heat is created due to friction between the shoulder and the workpiece, resulting in plastic deformation of the materials in the stir zone [100]. High strain and heat energies generated during the stirring of the tool on the workpiece cause dynamic recrystallization, which causes the formation of new grains in the weld zone/stir zone [101].

To maintain better weld quality, several parameters are responsible such as tool geometry, tool material, coating of the tool and welding parameters like rotational speed, travel speed, and plunge depth [102]. Heat is mainly induced by the tool shoulder rather than the pin [103]. In the case of welding hard materials like steel, the process has certain limitations because to weld hard materials a very highly durable tool is required, and a high temperature is also required which is difficult to produce by the tool pin and shoulder; welding speed cannot be attained as well as on aluminium alloys due to the high hardness of steels and high tool damage rate is very high [97]. However, few tool materials like commercial pure tungsten and polycrystalline cubic boron nitride (PCBN) tools can survive these problems, but it increases the welding cost by eight times or even more when compared to fusion welding [97]. The FSW tools for steel/hard materials are expensive even though this welding process has abundant advantages compared to conventional fusion welding processes, such as less HAZ formation, no filler requirements, it eliminates hydrogen embrittlement, porosity and several other defects over fusion welding [97,104]. Friction stir welds show more homogeneous grain structure and better mechanical properties and corrosion resistance compared to fusion welding [98,104].

2.2.2. Friction welding

Friction welding is another kind of pressure-based solid-state joining process [105] because weld is produced below the melting point of the materials [105,106].

Initially, one workpiece is rotated while the other is kept stationary as illustrated in Figure 1.8. When the desired rotational speed is accomplished, the two workpieces are brought into contact with one another and axial force is applied, as shown in Figure 1.7b, and due to rubbing action at the interface of the workpiece the heat initiates upsetting as shown in Figure 1.7c. Finally, rotation of one workpiece is stopped which completes the upsetting (Figure 1.7d). The weld formed is shown as narrow HAZ [107]. Friction welding has a several versions such as rotary friction welding, orbital friction welding and linear friction welding [107-110]. Rotary

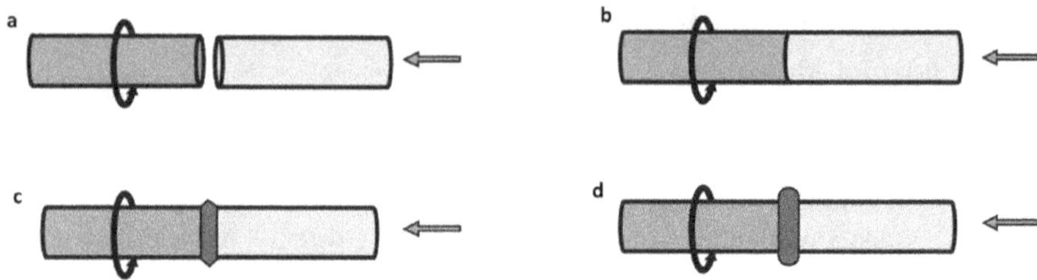

FIGURE 1.8 Steps in friction welding

friction welding has been developed and commercialized since the 1940s, in which one component is rotated around its axis while the other remains stationary; it is not useful to join non-circular cross-section parts because the rate of heat generation is not uniform in the interface [107,108]. However, these limitations can be averted by using orbital or linear friction welding. Orbital friction welding, which is a hybrid combination of linear and rotary friction welding, was developed and introduced in the 1970s. In this process, the center of one component relative to the other component is moved around a two-dimensional curve, for example a circle, to provide the rubbing action. The two parts to be joined are rotated around their longitudinal axes in the same manner with the same constant angular speed [107]. On the other hand, in linear friction welding, which was developed and commercialized in the 1980s, the components move under friction pressure relative to each other in a reciprocating manner so that a small linear displacement in the plane of the joint can be made [107, 110].

2.2.3. Ultrasonic welding

Ultrasonic welding (USW) is proven to have great advantages as green welding technology nowadays in connecting the electrical system of new energy vehicles [111]. Among solid-state joining, USW is the fastest and highly energy-efficient process [112]. For bonding composites and plastics, USW technology is frequently used in various engineering industries, especially in the automotive and textile industry, in the production of electrical appliances, and in the packaging technology [113].

Ultrasonic welding is gaining popularity as an efficient solution for thin metal components joining [114]. It has several advantages in comparison with RSW and FSW in terms of low-energy input [113]. An ultrasonic welding device is categorized into five fundamental parts, that is, power supply (for generating the high-frequency pulses), piezo transducer (converts high-frequency electrical energy into vibration or mechanical energy), wedge (amplify vibration amplification) and sonotrode (facilitate the uniform pressure and vibration toward the metal parts to be joined) and pneumatic cylinder (clamping pressure to the adjoining materials during joining [115]. During welding uses ultrasonic energy is generated at high frequencies around 20–40 kHz to produce mechanical vibrations at a low amplitude of around 1–25 μm and due to vibrations heat is generated at the joint interface of the welded parts, resulting in melting of the materials [116,117] as shown in Figure 1.9. It is the fastest known welding technique, with a welding time in between 0.1 and 1.0 seconds [117].

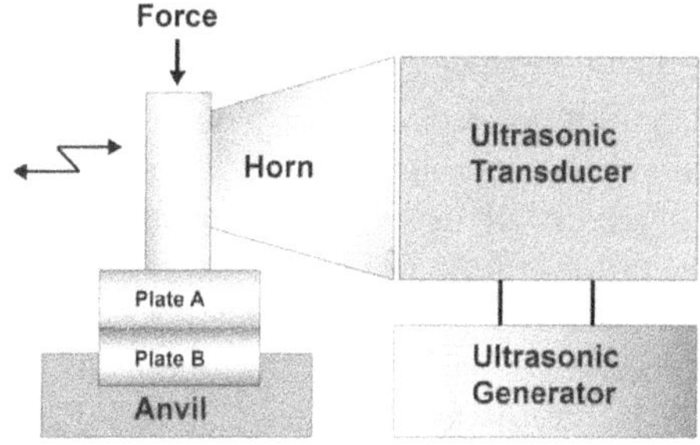

FIGURE 1.9 Ultrasonic welding

3. SCOPE AND APPLICATION OF WELDING

Welds are replacing rivets in various components in both commercial airplanes and the military to enhance both cost and structural integrity [118]. Diffusion welding, LBW and EBW are mostly preferred in commercial aircraft and fighter planes. Electron beam welding is continually rising in importance in military airplanes for the joining of titanium alloys [119]. Even for space technology, after a long discovery (in the 21st century), a vacuum EBW is adopted as suitable, reliable, versatile and efficient welding in space technology [120]. Laser beam welds completely replaced the mechanical rivets in large parts of the fuse in all commercial airplanes [121]. Some new processes developed for the space industry have already proven to be promising for the aeronautics industry [118,120], and these include FSW and variable polarity plasma arc welding (PAW), for critical applications in the rockets industry.

But still one process that does not gained widespread application is the diffusion welding of aluminum alloys which is the scope of further research.

The worldwide growth in the construction industry, automotive, aerospace, transport and shipbuilding industries is ensuing high growth in the market, especially in developing countries like India and China and other countries. The expanding demand is because of increased activities in various manufacturing sectors.

Welding and joining are hugely applied in many industries. The anticipated investments show the potential for welding applications.

In the construction industry the estimated investment in urban infrastructural projects is of US$ 650 billion. Total planned investment in the infrastructure sector is US$ one trillion [122].

In shipbuilding sectors welding has high growth prospects. Opportunities are assessed because half of the current fleet of ships is more than 20 years old, and this presents scope for large opportunities for growth in the shipbuilding and repair industries [60, 61, 124]. A lot of advantages are in railways: antirust, lightweight, and low cost of maintenance and a simple manufacturing process, meaning the production of railway vehicles has become the development strategy of European, American, and other developed nations [125,126]. The automobile industry mainly used arc welding, spot welding, projection welding, and many other joining methods, but for the consideration of work efficiency has recently emphasized the ease of one-side welding of closed section components and continuous welding work; this can be fulfilled by using laser welding and laser brazing in various automotive components [60,61,127].

There are huge opportunities for welding manufacturers. The enormous drive behind the changes in various approaches to innovation, adopting new technology and the courage to invest in all fields of manufacturing reflects manufacturers' commitment and tenacity toward the improvements of processes, for example integration and adoption of dissimilar materials in a larger number of components. This approach is basically observed in sectors like the automobile industry due to current trends and demands of the consumers; by considering the current use of low-level technologies, advantage can be taken and in jumping directly to the latest advanced welding technologies. The important factor in arc welding to be noted is that it will remain a widely used welding process in the near future, and this is true across the globe. Arc welding is about 35% of global welding from manual to automated processes [123,128].

However, there are several challenges in the welding industry. This industry is basically driven by various advanced welding methods and equipment and a wide range of scope of applications. Conversely, in the global scenario the scarcity of manpower is the principal influencing factor in the implementation of automation by using robots in almost all the manufacturing industries. Furthermore, implementation of the modern advanced welding techniques and associated processes such as controlled joining, feedback systems with sensors are helping to increase productivity and flexibility.

4. FUTURE OPPORTUNITY IN WELDING

In the last several decades, welding has evolved as an interdisciplinary activity due to the huge industrial revolution and continues to evolve new materials; the technology of welding creates a lot of opportunities in various sectors or laboratory/commercial scale. To find out the suitability and/or compatibility of a process with materials in terms of physical, chemical, mechanical, or metallurgical complication, selection of filler materials is still challenging. The joining of aluminium alloys with steels shows difficulties due to the metallurgical incompatibility between these two materials [129]. Space technology is looking for the suitability of welding processes, however, welding processes still need to be established. The application of diffusion welding in aluminium alloy still opens the scope. High nitrogen-based steels which show excellent tensile strength and impact properties due to the solid solution strengthening of nitrogen, still show difficulties in welding due to being prone to spatters, the severe loss of nitrogen, and the poor performance of weld joints [130]. Aluminium 6xxx grade extrusions which are more popular in the automotive industry for manufacturing body-in-white and battery tray remains susceptible to the weld cracking [131] which opens the possibility for further improvement in and technique to weld. Joining shape memory and high entropy alloys still opens the scope for research.

Underwater welding shows major issues because the welded workpiece immediately comes into contact with water, leading to a severe quenching effect resulting in hydrogen diffusion that causes hydrogen-induced cracking, arc instability, and weld porosity [128]; it is still challenging to find better welding conditions and processes, therefore opening opportunity for further research.

5. CONCLUSION

The article aims to review and collectively uncover the welding approaches which have been adopted for various applications and the reason behind the lack of popularity in industries for mass adoption. Thus, the following conclusions have been made:

1) There is lot of scope and opportunity for welding processes in the fields of space technology, aerospace, automobile, shipbuilding, marine, oil and military.
2) To meet the worldwide huge competition and survive the products in the global market, a new way of thinking is inevitable to change and continue to improve the existing welding technology and provide products at economical cost.
3) Conversely, in the global scenario the shortage of manpower is the primary influencing factor in the implementation of automation by using robots in almost all the manufacturing industries.
4) The developments of advanced and sophisticated welding techniques captured the market and completely replaced rivets.
5) Electron beam welding is continually adopted for the joining of titanium alloy in military airplanes.
6) Modified versions of TIG/GTAW maintain sustainability and improve weld penetration up to 1.5–4 times in a single pass itself.
7) Development of 1-micron wavelength laser welding technology over 10-micron wavelength shows better weld quality and greater flexibility.
8) AI and robotized welding techniques are truly remarkable today as one of the major application areas for the manufacturing industry as they improve productivity in a shorter time.

REFERENCES

[1] E. R. Bohnart, *Welding: Principles and Practices*, 5th edn. New York: McGraw-Hill Education, 2017.
[2] H. Soltan, M. Omar, A roadmap for selection of metal welding process: A review and proposals, *Welding in the World*, 66 (2022).
[3] Market Research Report, www.fortunebusinessinsights.com/industry- reports/welding-market-101657, 2023.
[4] O. K. Makovetskaya, Modern market of welding equipment and materials, *Middle East*, 16 (2011), 17–2.
[5] T. S. Hong, M. Ghobakhloo, W. Khaksar, Robotic welding technology, *Comprehensive Materials Processing*, 6 (2014).
[6] P. Kah, J. Martikainen, Curret trends in welding processes and materials: Improve in effectiveness, *Reviews on Advanced Materials Science*, 30 (2012).
[7] R. G. Narayanan, J. S. Gunasekera, Introduction to sustainable manufacturing processes, in *Sustainable Manufacturing Processes*, 2023.

[8] H. Rana, V. Badheka, P. Patel, V. Patel, W. Li, J. Andersson, Augmentation of weld penetration by flux assisted TIG welding and its distinct variants for oxygen free copper, *Journal of Materials Research and Technology*, 10 (2021).

[9] S. R. Singh, P. Khanna, A-TIG (activated flux tungsten inert gas) welding A review, *Materials Today: Proceedings*, 44 (2021).

[10] A. V. S. Babu, P. K. Giridharan, P. R. Narayanan, S. V. S. Narayana Murty, Prediction of bead geometry for flux bounded TIG welding of AA 2219-T87 aluminum alloy, *Journal of Advanced Manufacturing Systems*, 15 (2016).

[11] J. Sivakumar, N. B. Karthik Babu, M. P. Mohanraj, E. Hariharan, M. Ranjithkumar, A new perception of Activated Flux Tungsten Inert Gas (A-TIG) *Welding Techniques for Various Materials*, 5 (2022).

[12] A. E M. Omer, et al., A review on friction stir welding of thermoplastic materials: Recent advances and progress, *Welding in the World* (2022).

[13] R. Kumar, , R. Singh, I. P. S. Ahuja, R. Penna, L. Feo, Weldability of thermoplastic materials for friction stir welding-A state of art review and future applications, *Composites Part B: Engineering*, 137 (2018).

[14] R. Saha, P. Biswas. "Current status and development of external energy-assisted friction stir welding processes: A review." *Welding in the World*, 66.3 (2022): 577–609.

[15] G. K. Padhy, C. S. Wu, S. Gao, Auxiliary energy assisted friction stir welding-status review, *Science and Technology of Welding and Joining*, 20 (2015).

[16] A. I. Alvarez, ´ M. García, G. Pena, J. Sotelo, D. Verdera, Evaluation of an induction assisted friction stir welding technique for super duplex stainless steels, *Surface and Interface Analysis*, 46 (2014) 892–896, https://doi.org/10.1002/sia.5442

[17] S. Muthukumaran, A process for friction welding of tube to tube sheet or a plate by adopting an external tool. *Indian patent patent no. 217446*, 2008.

[18] S. Kumaran, S. S. Muthukumaran, D. Venkateswarlu, G. K. Balaji, S. Vinodh, Eco-friendly aspects associated with friction welding of tube-to-tube plate using an external tool process, *International Journal of Sustainable Engineering* 5 (2012).

[19] R. G. Narayanan, J. S. Gunasekera, Introduction to sustainable manufacturing processes, *Sustainable Manufacturing Processes*, Academic Press, 2023.

[20] S. B. Chen, N. Lv., Research evolution on intelligentized technologies for arc welding process, *Journal of Manufacturing Processes*, 16.1 (2014).

[21] G. Bolmsjo, M. Olsson, P. Cederberg, Robotic arc welding–trends and developments for higher autonomy, *Industrial Robot: An International Journal* (2002).

[22] Q. Liu, C. Chen, S. Chen, Key technology of intelligentized welding manufacturing and systems based on the Internet of Things and multi-agent, *Journal of Manufacturing and Materials Processing*, 6 (2022).

[23] S. B. Chen, On intelligentized welding manufacturing, In *International Conference on Robotic Welding, Intelligence and Automation*. Cham: Springer International Publishing, 2014.

[24] R. Tsuzuki, Development of automation and artificial intelligence technology for welding and inspection process in aircraft industry, *Welding in the World*, 66.1 (2022).

[25] H. Wang, Y. Wang, *High-Velocity Impact Welding Process: A Review*, Metals, (2019).

[26] Y. Zhang, S. S. Babu, C. Prothe, M. Blakely, et al., Application of high velocity impact welding at varied different length scales, *Journal of Materials Processing Technology*, 211 (2011).

[27] T. Patterson, et al., A review of high energy density beam processes for welding and additive manufacturing applications, *Welding in the World*, 65.7 (2021).

[28] T. Omi, K. Numano, The role of the CO2 laser and fractional CO2 laser in dermatology, *Laser Therapy*, 23 (2014).
[29] M. Węglowski, S. Błacha, A. Phillips, Electron beam welding-Techniques and trends-Review, *Vacuum*, 130 (2016).
[30] K. A. Srivastava, S. Ashutosh, Advances in joining and welding technologies for automotive and electronic applications, *American Journal of Materials Engineering and Technology*, 5 (2017).
[31] Welding handbook, 8th edition, volume 1 & 2, American Welding Society, USA 1987.
[32] J. M. Antonini, *Health Effects Associated with Welding*, 2014.
[33] R. Singh, *Applied Welding Engineering: Processes, Codes, and Standards*. Butterworth-Heinemann, 2020.
[34] G. Caballero, Francisca, M. W. Fu, et al., Encyclopedia of Materials: Metals and Alloys, (2022).
[35] M. Hashmi, J. Saleem. *Comprehensive Materials Processing*. Newnes, 2014.
[36] K. Sham, S. Liu, Flux coating development for SMAW consumable electrode of high nickel alloys, *Welding Journal*, 93 (2014).
[37] P. P. Thakur, A. N. Chapgaon, A review on effects of GTAW process parameters on weld, *IJRASET*, 4 (2016).
[38] D. W Rathod, Comprehensive analysis of gas tungsten arc welding technique for Ni-base weld overlay, *Advanced Welding and Deforming*, (2021).
[39] F. Pichot, et al., Numerical definition of an equivalent GTAW heat source, *Journal of Materials Processing Technology*, 213.7 (2013).
[40] T. A. Tejedor, R. Singh, P. Pilidis, *Maintenance and Repair of Gas Turbine Components, Modern Gas Turbine Systems*. Woodhead Publishing, 2013.
[41] P. Jansohn, ed., Modern gas turbine systems: High efficiency, low emission, fuel flexible power generation, (2013).
[42] D. H. Phillips, *Welding Engineering: An Introduction*. John Wiley & Sons, 2023.
[43] H. Arora, R. Singh, G. S. Brar, Thermal and structural modelling of arc welding processes: A literature review, *Measurement and Control*, 52 (2019).
[44] A. Sarolkar, K. Kolhe, A Review of (GTAW) Gas Tungsten Arc Welding and its parameters for joining aluminium alloy, *International Journal for Science and Advance Research in Technology*, 3 (2017).
[45] R. S. Vidyarthy, and D. K. Dwivedi. Activating flux tungsten inert gas welding for enhanced weld penetration. Journal of Manufacturing Processes, 22 (2016) 211–228.
[46] K. Kumar et al, A review on TIG welding technology variants and its effect on weld geometry, *Materials Today: Proceedings*, 50 (2022).
[47] L. D. Smith, Gas tungsten arc welding fundamentals: Understanding GTAW, (2013). www.thefabricator.com/thewelder/article/arcwelding/gas-tungsten-arc-welding- fundamentals-understanding-gtaw
[48] D. K. Dwivedi, Arc welding processes: Gas Tungsten Arc Welding: Electrode, polarity and pulse variant, fundamentals of metal joining: Processes, *Mechanism and Performance*, (2022).
[49] Specifications, *Standard Welding Procedure, ASM handbook, volume 6A: Welding Fundamentals and Processes*. 2011.
[50] K. Weman, *Welding Processes Handbook*. Elsevier, 2011.
[51] I. A. Ibrahim et al, The effect of Gas Metal Arc Welding (GMAW) processes on different welding parameters, *Procedia Engineering*, 41 (2012).
[52] J. Norrish, *Gas Metal Arc Welding, Advanced Welding Processes*, (2006) 100–135, https://doi.org/10.1533/9781845691707.100

[53] H. Kumar, N. Kumar, Manmohan, *A Survey On Gas Metal Arc Welding* (GMAW)-Review, (2019).

[54] K. H. Li, J. S. Chen, Yu. M. Zhang, Double-electrode GMAW process and control, *Welding Journal-New York-* 86, (2007).

[55] J. H. Waszink, G. J. P. M. Van den Heuvel, *Heat Generation and Heat Flow in the Filler Metal in GMA Welding*, 1982.

[56] J. Norrish, D. Cuiuri, The controlled short circuit GMAW process: A tutorial, *Journal of Manufacturing Processes*, 16 (2014).

[57] S. Tsushima, M. Kitamura, Tandem electrode AC-MIG welding-development of AC-MIG welding process (Report 4), *Welding Research Abroad*, 42 (1996).

[58] V. P. Dinbandhu, J. J. Vora, K. Abhishek, Advances in gas metal arc welding process: Modifications in short-circuiting transfer mode, *Advanced Welding and Deforming*. Amsterdam, The Netherlands, Elsevier, 2021.

[59] S. Vyas, S. Shah, G. D. Acharya, *Minimization of Distortion During Gas Metal Arc Welding Process used for Hydraulic Shearing Machine Pressure Plate Square*, 2018.

[60] H. Mallaradhya, K. M. Vijay, R. Ranganatha, S. Darshan, Resistance spot welding: A review, *International Journal of Mechanical and Production Engineering Research and Development*, 8 (2018).

[61] H. Zhang, J. Senkara, *Resistance Welding: Fundamentals and Applications*. CRC press, 2011.

[62] P. A. Dhawale, M. L. Kulkarni, Electric resistance spot welding: A state of art, *International Journal of Engineering Research & Technology (IJERT)*, 5, (2017).

[63] A. S. Bharaj, S. Shukla, S. Gedam, R. Kukde, S. Verulkar, Study of resistant spot welding and its effect on the metallurgical and mechanical properties a review, *Materials Today: Proceedings* (2023).

[64] K.. Weman, *Welding Processes Handbook*. Elsevier, 2011.

[65] E. P. Degarmo, J. T. Black, R. A. Kohser, *Materials and Processes in Manufacturing*, 1997.

[66] Y. Ma, P. Wu, et al., Review on techniques for on-line monitoring of resistance spot welding process, *Advances in Materials Science and Engineering* (2013).

[67] R. W. Messler, *Integral Mechanical Attachment: A Resurgence of the Oldest Method of Joining*. Elsevier, 2011.

[68] R. P. Reddy, Analysis of spot welding electrode, *International Journal of Mechanical and Industrial Engineering (JIMIE)*, 2, (2012).

[69] R. P. Singh, S. Kumar, S. Dubey, A. Singh, A review on working and applications of oxy-acetylene gas welding, *Materials Today: Proceedings*, 38 (2021).

[70] J. G. Buijnsters, F. M. Van Bouwelen, J. J. Schermer, W. J. P. Van Enckevort, Chemical vapour deposition of diamond on nitrided chromium using an oxyacetylene flame." *Diamond and Related Materials*, 9 (2000).

[71] A. P. Kulkarni, P. Randive, A. R. Mache, Micro-controller based oxy-fuel profile cutting system." *International Journal of Mechanical and Mechatronics Engineering*, 2 (2008).

[72] S. A. Civjan, T. Guihan, K. Peterman, Testing of oxyacetylene weld strength, *Journal of Constructional Steel Research*, 168 (2020).

[73] B. D.. Richardson, T. Z. Blanzymski, E. N. Gregory, et al., Manufacturing methods, In *Mechanical Engineer's Reference Book*, pp. 16–1. Butterworth-Heinemann, 1994.

[74] J. Thamilarasan, N. Karunakaran, P. Nandakumar, Optimization of oxyacetylene flame hardening parameters to analysis the surface structure of low carbon steel, *Materials Today: Proceedings*, 46 (2021).

[75] K. Hokamoto, ed. *Explosion, Shockwave and High-Strain-Rate Phenomena of Advanced Materials*. Elsevier, 2021.
[76] S. T. Mileiko, *Metal and Ceramic Based Composites*. Elsevier, 1997.
[77] B. B. Sherpa, D. K. B. Pal, U. Batra, G. C. Mishra, B. B. Singh, Study of the explosive welding process and applications, *Advances in Applied Physical and Chemical Sciences-a Sustainable Approach*, (2014).
[78] F. T. Mahi, U. Dilthey. *Joining of metals*, (2016).
[79] G. H. S. F. L. Carvalho, et al., The role of physical properties in explosive welding of copper to stainless steel, *Defence Technology*, 22 (2023).
[80] J. B. Ribeiro, R. Mendes, A. Loureiro, Review of the weldability window concept and equations for explosive welding, In *Journal of Physics: Conference Series*, IOP Publishing, (2014).
[81] R. Timings, *Fabrication and Welding Engineering*. Routledge, 2008.
[82] H. Z. Oo, P. Muangjunburee, Improving microstructure and hardness of softening area at HAZ of thermite welding on rail running surface, *Materials Today Communications* 34 (2023).
[83] H. B. Yun, K. C. Lee, Y. J. Park, D. K. Jung, Rail neutral temperature monitoring using non-contact photoluminescence Piezospectroscopy: A field study at high-speed rail track, *Construction and Building Materials*, 204 (2019). https://doi.org/10.1016/j.conbuildmat.2019.01.199
[84] Welding handbook. Volume 3, Fl Welding Processes, Part 2, American Welding Society, Miami, 2007.
[85] C. P. Lonsdale, M. Engineer. Thermite rail welding: History, process developments, current practices and outlook for the 21st century. In *Proceedings of AREMA 1999 Annual Conference*, 1895 (1999).
[86] J. R. Deepak, R. P. Anirudh, S. S. Sundar, Applications of lasers in industries and laser welding: A review, *Materials Today: Proceedings* (2023).
[87] H. Lee, C. H. J. Lim, M. J. Low, N. Tham, et al., Lasers in additive manufacturing: A review, *International Journal of Precision Engineering and Manufacturing-Green Technology*, 4 (2017).
[88] D. Wallerstein, A. Salminen, F. Lusquiños, et al., Recent developments in laser welding of aluminum alloys to steel, *Metals*, 11 (2021).
[89] B. M. X. Gao, Y. Huang, et al., A review of laser welding for aluminium and copper dissimilar metals, *Optics & Laser Technology*, 167 (2023).
[90] S. Katayama, ed. *Handbook of Laser Welding Technologies*. Elsevier, 2013.
[91] C. E. Webb, J. D. C. Jones (Eds.), *Handbook of Laser Technology and Applications: Laser Design and Laser Systems*. CRC Press, 2004.
[92] W. Ke, Z. Zeng, J. P. Oliveira, B. Peng et al., Heat transfer and melt flow of keyhole, transition and conduction modes in laser beam oscillating welding, *International Journal of Heat and Mass Transfer*, 203 (2023).
[93] J. Svenungsson, I. Choquet, A. F. H Kaplan, Laser welding process–a review of keyhole welding modelling, *Physics Procedia*, 78 (2015).
[94] M. St. Węglowski, S. Błacha, A. Phillips, Electron beam welding-Techniques and trends-Review, *Vacuum*, 130 (2016).
[95] E. V. Walle, *Mechanical Test Specimens, Reconstitution of Encyclopedia of Materials: Science and Technology*, 2001.
[96] Z. Sun, R. Karppi, The application of electron beam welding for the joining of dissimilar metals: An overview, *Journal of Materials Processing Technology*, (1996).
[97] G. D. Mohan, C. S. Wu, A review on friction stir welding of steels, *Chinese Journal of Mechanical Engineering*, 34 (2021).

[98] F. C. Liu, Y. Hovanski, M. P. Miles, C. D. Sorensen, T. W. Nelson, A review of friction stir welding of steels: Tool, material flow, microstructure, and properties, *Journal of Materials Science & Technology*, 34 (2018).

[99] R. V. Arunprasad, G. Surendhiran, M. Ragul, T. Soundarrajan, S. Moutheepan, S. Boopathi., Review on friction stir welding process, *International Journal of Applied Engineering. Res. ISSN*, 13 (2018).

[100] V. P. Singh, et al., Recent research progress in solid state friction-stir welding of aluminium-magnesium alloys: A critical review, *Journal of Materials Research and Technology*, 9 (2020).

[101] P. K. Sahu, S. Pal, Influence of metallic foil alloying by FSW process on mechanical properties and metallurgical characterization of AM20 Mg alloy, *Materials Science and Engineering*, 684 (2017).

[102] V. K. Mahakur, K. Gouda, P. K. Patowari, S. Bhowmik, A review on advancement in friction stir welding considering the tool and material parameters." *Arabian Journal for Science and Engineering*, 46 (2021).

[103] D. Jacquin, G. A. Guillemot, Review of microstructural changes occurring during FSW in aluminium alloys and their modelling, *Journal of Materials Processing Technology*, (2021).

[104] D. G. Mohan, S. Gopi, Infuence of In-situ induction heated friction stir welding on tensile, microhardness, corrosion resistance and microstructural properties of martensitic steel, *Engineering Research Express*, (2021).

[105] R. Paventhan, P. R. Lakshminarayanan, V. Balasubramanian, Optimization of friction welding process parameters for joining carbon steel and stainless steel, *Journal of Iron and Steel Research International*, 19 (2012).

[106] A. Ramesh, M. Pranav, P. Subramaniyan, Eswaran, Review on friction welding of similar/dissimilar metals, In *Journal of Physics: Conference Series*, (2019).

[107] M. B. Uday, M. N. A. Fauzi, H. Zuhailawati, A. B. Ismail, Advances in friction welding process: A review, *Science and Technology of Welding and Joining*, 15 (2010).

[108] R. Selvaraj, K. Shanmugam, P. Selvaraj, V. Balasubramanian, Optimization of process parameters of rotary friction welding of low alloy steel tubes using response surface methodology, *Forces in Mechanics*, 10 (2023).

[109] U. Raab, S. Levin, L. Wagner, C. Heinze, Orbital friction welding as an alternative process for blisk manufacturing, *Journal of Materials Processing Technology*, 215 (2015).

[110] M. Javidikia, M. Sadeghifar, H. Champliaud, M. Jahazi, Grain size and temperature evolutions during linear friction welding of Ni-base superalloy Waspaloy, Simulations and experimental validations, *Journal of Advanced Joining Processes*, (2023).

[111] X. M. Cheng, K. Yang, J. Wang, W. T. Xiao, S. S. Huang, Ultrasonic system and ultrasonic metal welding performance: A status review, *Journal of Manufacturing Processes*, 84 (2022).

[112] T. G. Unnikrishnan, P. Kavan, A review study in ultrasonic-welding of similar and dissimilar thermoplastic polymers and its composites, *Materials Today: Proceedings*, 56 (2022).

[113] T. Nguyenvo, P. Lenfeld, A review of studies on ultrasonic welding, *Annals of the Faculty of Engineering Hunedoara*, 15 (2017).

[114] Z. L. Ni, F. X. Ye, Ultrasonic spot welding of aluminum alloys: A review, *Journal of Manufacturing Processes*, 35 (2018).

[115] F. Rubino, H. Parmar, et al., Ultrasonic welding of magnesium alloys: A review, *Materials and Manufacturing Processes*, 35 (2020).

[116] H. Potente, Ultrasonic welding-Principles & theory, *Materials & Design*, 5 (1984).

[117] J. M. Troughton, *Handbook of Plastics Joining: A Practical Guide*. William Andrew, 2008.

[118] F. P. Mendez, et al., Welding processes for aeronautics, *Advanced Materials and Processes*, 159 (2001).

[119] B. Choudhury, M. Chandrasekaran, Investigation on welding characteristics of aerospace materials-A review, *Materials Today: Proceedings*, 4 (2017).

[120] W. Ting, W. Yifan, G. Chong, J. Siyuan, Feasibility study on feeding wire electron beam brazing of pure titanium using an electron gun for space welding, *Vacuum*, 180 (2020).

[121] Y. Chen, L. Li, W. Tao, Z. Yang, Laser welding technologies for aircraft fuselage panels, In *Laser and Tera-Hertz Science and Technology*, Optica Publishing Group, 2012.

[122] M. Mazur, Assessment of the construction welding process, *Procedia Engineering*, 192 (2017).

[123] Advanced Welding Technologies for Fabrication Industry, 2020 www.dcmsme.gov.in/white_paper/7.%20Advance%20Welding%20Technologies% 20for%20 Fabrication%20Industry-Year%202.pdf

[124] E. Turan, T. Kocal, et al., Welding technologies in shipbuilding industry, *Tojsat*, 1 (2011).

[125] H. Xiao, et al., Analysis on mechanical characteristics of welded joint with a new reinforced device in high-speed railway, *Advances in Mechanical Engineering*, 12 (2020).

[126] Steenbergen, M. J. M. M., R. W. V. Bezooijen, Rail welds, In *Wheel-Rail Interface Handbook*, Woodhead Publishing, 2009.

[127] D. Devarasiddappa, Automotive applications of welding technology-a study, *International Journal of Modern Engineering Research*, 4 (2014).

[128] P. Vashishtha, R. Wattal, S. Pandey, N. Bhadauria, Problems encountered in underwater welding and remedies-a review, *Materials Today: Proceedings*, 64 (2022).

[129] C. Lucian, I. Mitelea, et al., The opportunity to join through laser braze-welding of aluminium alloys with galvanized low alloy steels. *Materials Today: Proceedings*, (2021).

[130] L. Wang, Y. Li, et al., Problems in welding of high nitrogen steel: A review, *Metals*, 12 (2022).

[131] T. Sun, et al., Challenges and opportunities in laser welding of 6xxx high strength aluminium extrusions in automotive battery tray construction, *Procedia CIRP*, 94 (2020).

2 Various Advanced Welding Methods: A Brief Overview

Gaurav

1. INTRODUCTION

The conventional welding process involves the joining of two or more materials by applying heat to the surfaces to the melting point and/or without the application of pressure or filler material. It allows the materials to cool and solidify, forming a strong bond between the joints. It can be done by various techniques that include arc welding, gas welding, and resistance welding. It has been a long-established method of manufacturing products, from consumer products to heavy machinery, for many years.

The welding technique has evolved over the years, and new, advanced methods have been developed to meet the needs of different industries. The term "advanced welding methods" refers to techniques that go beyond traditional welding. For instance, a major difference between traditional welding and advanced welding methods is the amount of heat that is used. High temperatures are typically used in traditional welding methods, resulting in distortion and damage to the metal being welded. In contrast, advanced welding methods operate at lower temperatures and can join a wider range of materials, including dissimilar metals. In general, advanced welding methods are better than traditional welding methods in terms of precision and control, heat input, speeds, and the ability to join a wider range of materials. Some examples of advanced welding techniques include laser welding, electron beam welding, hybrid welding, friction stir welding, robotic welding, and explosive welding. In addition to expanding the range of materials that can be welded, these methods have enabled industries to create new and innovative products and processes. The purpose of this chapter is to provide an overview of various advanced welding methods that have been developed in recent years. We can expect even more advanced welding methods in the years to come as a result of ongoing research and development.

2. VARIOUS ADVANCED WELDING

2.1. Laser Welding

Laser welding or laser beam welding (LBW) is one of the modern and innovative techniques of joining materials, usually metals or thermoplastics, using a laser beam

DOI: 10.1201/9781003435884-2

TABLE 2.1
Comparison of LBW and EBW

Parameters	LBW	EBW
Beam source	LBW uses a beam of photons generated by a laser source and amplified by optical devices.	EBW uses a beam of electrons generated by an electron gun and accelerated by electric fields.
Beam environment	LBW is not performed under a vacuum environment and usually requires a shielding gas such as argon or nitrogen to protect the weld from oxidation.	EBW is performed in a high vacuum environment, which prevents the beam from scattering by gas molecules and eliminates impurities from the weld zone.
Beam penetration	LBW can produce shallower and wider welds than EBW, as photons have lower kinetic energy and higher divergence than electrons.	EBW can produce deeper and narrower welds than LBW, as electrons have higher kinetic energy and lower divergence than photons.
Beam safety	LBW can also produce harmful radiation, but less than EBW, and does not produce vapors.	EBW can produce harmful radiation and vapors that need to be shielded and vented properly.
Beam cost	LBW also requires expensive equipment, but less than EBW, and can be more easily automated.	EBW requires expensive and complex equipment and maintenance, as well as skilled operators and technicians.
Filler material	It can weld metals with or without filler material, depending on the joint design and gap.	The electron beam can weld metals without filler material, creating a clean and strong weld.
Types of materials to be welded	It can weld different types of metals, including dissimilar metals, alloys, and coated metals.	It can weld metals with high melting points, high strength, or high hardness, such as titanium, refractory metals, and flammable metals.
Direction of weld	The laser beam can weld metals in any position or orientation, due to the flexibility of the laser source and delivery system.	The electron beam can weld metals in any position or orientation, due to the flexibility of the magnetic lens and worktable.

as the radiant energy source. This welding process is non-contact, that is, does not require pressure and utilizes inert gas to prevent oxidation of the molten pool. Filler metals may also be used in some cases [1].

2.1.1. Working principle

A schematic diagram of the LBW setup is shown in Figure 2.1. The welding procedure has been divided into simple steps.

TABLE 2.2
Advantages and disadvantages of FSW

	Advantages	Disadvantages
1.	High-quality welds with minimal distortion and defects.	High equipment cost and power consumption.
2.	It does not require consumables, shielding gas, flux, or filler material.	It creates a visible hole in the welding plates where the tool exits.
3.	Join dissimilar metals and alloys that are difficult to weld by other methods.	Limited to butt or lap joints with simple geometries.
4.	Less residual stresses and improves the fatigue resistance of the welds.	It cannot weld non-forgeable materials, such as cast iron, ceramics, or composites.
5.	Good repeatability and automation.	Typically used on materials with a low melting point, such as aluminum and certain plastics.
6.	Reduced environmental impact.	It may cause distortion of the workpieces due to the high frictional forces and thermal expansion.

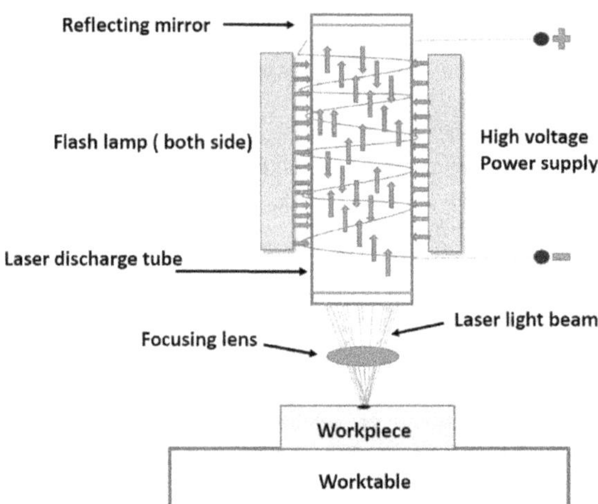

FIGURE 2.1 Schematic diagram of LBW setup

- **Machine location**: The welding equipment is initially set up in the desired area (between the two metal parts to be joined).
- **Production of concentrated laser beam**: The applied high-voltage power supply starts a flash lamp of the machine, which emits the light photons. The energy of light photons is absorbed by a laser beam medium (gas/solid-state/fiber laser), which causes the electrons to attain a higher energy level. When they return to their lower energy state, high-energy photons are produced. This

photon again stimulates the excited state of the atom of the laser beam medium and produces more photons. In this way, the process continues and we get a concentrated beam of light which is called LASER beam light. The term LASER stands for "Light Amplification by Stimulated Emission of Radiation" [2, 3].

- **CAM control**: It is utilized to control the laser machine and workpiece table during welding.
- **Melting and solidification**: A laser beam is used to heat and melt the joint between workpieces. Afterward, a weld is produced as the molten material solidifies along the path of the laser beam.

2.1.2. Modes of operation

Laser welding is essentially based on two modes of operation [2]. Conduction mode and keyhole mode are different operational regimes of laser welding that depend on the power density applied to the welding area. In conduction mode, the laser beam acts as a point source of energy whereas in keyhole mode, the beam acts as a line source of energy. Conduction mode takes place when the power intensity is not sufficient to cause boiling. The laser beam only gets absorbed at the surface of the material and does not penetrate it. Conduction-limited welds have a high ratio of width to depth while keyhole mode occurs when the intensity is high enough to cause vaporization and create a keyhole in the melt pool. In keyhole mode, deep penetration welding is achieved by moving the keyhole along the joint or the joint with respect to the laser beam, resulting in welds with a high depth-to-width ratio.

The laser beam can achieve a mixed mode of conduction and keyhole by adjusting power density, feed speed, and focusing. The welding can start with conduction mode and switch to keyhole mode by increasing power density or vice versa. Using a combination of conduction mode and keyhole mode in laser welding can provide several benefits, including better weld quality, reduced spatter, increased welding speed, and enhanced control over the weld geometry. However, this approach demands meticulous optimization of process parameters and more advanced equipment to attain the desired outcomes.

2.1.3. Types of laser

There are different types of lasers used in laser welding. A suitable laser source can be chosen for a welding project that depends on several factors, such as the material, thickness, shape, and quality of the parts to be welded, the desired welding speed and depth, the available budget and space, and the level of automation and flexibility required. Some of the common types of lasers used in laser welding are [2]:

Pulsed laser – Pulsed lasers deliver short bursts of high-energy pulses that create a series of overlapping spot welds along the joint. It can weld thin materials or join dissimilar materials with different melting points. It can also be suitable for microwelding applications that require high precision and minimal heat input.

Continuous laser – Continuous lasers deliver a constant beam of high-power density that creates a deep and narrow weld along the joint. It can weld thick materials or join similar materials with high melting points.

Gas laser – There are three types of gases used in these lasers: helium, nitrogen, and carbon dioxide. The laser beam is produced by exciting a gas mixture with a low-current, high-voltage power source. They can operate in a pulsed or continuous mode. It is suitable for welding nonmetallic materials such as plastics and glass, as well as metallic materials with low reflectivity such as steel and titanium.

Solid-state laser – It produces a laser beam by doping a solid medium such as crystal, glass, or fiber with rare-earth elements like neodymium or ytterbium [4]. It may weld to metallic materials with high reflectivity (such as aluminum or copper) and refractory materials such as quartz or ceramics. Solid-state lasers are also suitable for welding small parts or intricate shapes that require a short focal length. Solid-state lasers produce wavelengths of 1.06 μm, whereas CO_2 lasers produce 10.6 μm wavelengths.

Fiber laser – Fiber lasers use an optical fiber doped with rare-earth elements as the active medium and can be coupled with other fibers to deliver the laser beam to the workpiece. Fiber lasers can also be easily integrated with other systems, such as robots, CNC machines, or vision systems, to achieve high levels of automation and accuracy. It can be used for more flexibility and versatility in the welding process.

Therefore, the type of laser source used for welding can affect the choice of conduction or keyhole mode. Pulsed, gas, solid-state, and fiber lasers can be used for both conduction and keyhole mode welding, depending on the power density, pulse length, and beam diameter settings. Continuous lasers are more likely to be used for keyhole mode welding only.

2.1.4. Advantages and disadvantages of laser welding

It provides quick welding speed, deep welding depth, and minimal distortion. It can be used for welding at ambient temperature or under certain ambient conditions with simple welding equipment. It successfully joins dissimilar materials and can weld refractory materials such as titanium and quartz. It can achieve micro-welding by obtaining tiny spots, which can be accurately located. It is suitable for assembling and welding micro and small workpieces produced in large quantities. It can provide precise and localized heat input with minimal thermal effects. It can produce fine and aesthetic welds with minimal damage to the materials.

One of the main drawbacks is the high initial investment required for the laser and its associated systems, which may not be feasible for some low-volume or small-scale applications. Another challenge is the low gap tolerance, which means that the parts to be joined must be aligned and fitted very accurately. Laser welding also poses some health and safety risks, such as exposure to intense radiation, noise, and fumes from the laser beam. Therefore, the operators need to have specialized training and skills to use the laser equipment safely and effectively. Moreover, laser welding may not be compatible with all materials, as some of them may not absorb the laser beam well or may react negatively to the heat input. For instance, this process is not recommended for welding of high-carbon steels due to the high cooling rate that leads to the formation of cracks.

2.1.5. Applications

Laser welding is a versatile and efficient technique that can join or repair different materials (steels, HSLA steels, stainless steel, aluminum, and titanium) in various industries. It can create strong and narrow welds for large metal structures in shipbuilding, or lightweight and durable welds for car parts in the automobile industry. Moreover, it can ensure high-quality, biocompatibility, and hygiene for medical devices or implants in the medical industry.

2.1.6. Common defects in laser welding and its remedies [3]

Crack: Cracks that occur during laser continuous welding are mainly thermal cracks. The primary cause of these cracks is the large shrinkage force generated by the weld before it has fully solidified. To avoid these cracks, use wire filling, preheating, or other methods to reduce the shrinkage stress.

Air hole: Air holes or porosity are a common defect in laser welding caused by gas trapped in the fast-cooling molten pool. To avoid these pores, the workpiece surface should be cleaned, the blowing direction should be adjusted, and for aluminum alloys, a low-hydrogen filler metal or preheating should be used.

Splash: Splash or spatter is produced during laser welding and can significantly impact the surface quality of the weld and cause contamination and damage to the lens. To avoid spatter, you can reduce the welding energy or power density to avoid excessive vaporization of the material.

Undercut: Undercut is a defect that reduces the strength of the joint when the edge of the weld has a concave shape. To prevent undercut, a balanced heat input and output can be achieved by properly matching the power and speed of the laser beam. A filler metal or wire can also be used to add material at the edge of the weld.

Collapse: A collapse is caused by a convex weld cross-section at the joint edge, which increases the stress concentration and reduces the fatigue resistance. As molten pool size is reduced and excessive sagging of the material is prevented, welding speed or laser beam power can be increased to prevent collapse.

2.2. Electron Beam Welding

Electron beam welding (EBW) is a fusion welding process in which a high-speed electron beam causes two or more metal parts to melt and fuse together. A key principle of EBW is the conversion of electrical energy into kinetic energy of electrons, which is then converted into thermal energy at the workpiece surface [1-3].

Working principle: A high voltage is formed between the anode and cathode of the electron gun during this operation, which emits a stream of high-velocity electrons due to thermionic emission [4]. This electron beam is then focused on a small spot on the workpiece surface using a magnetic lens. Fixtures are used to clamp the workpieces

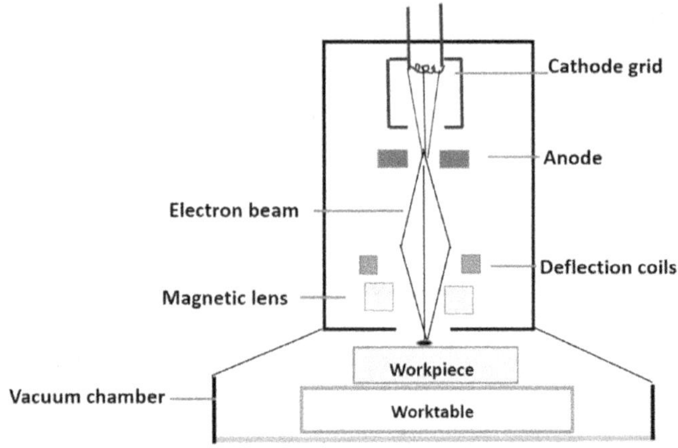

FIGURE 2.2 Schematic diagram of EBW setup

inside a vacuum chamber on a CNC-controlled worktable. As the worktable moves along the joint line, the workpieces melt, and the melted metal flows together to form a weld. A schematic diagram of the EBW setup is shown in Figure 2,2.

2.2.1. Advantages

- High depth-to-width ratio, which allows for welding thick sections with minimal distortion and residual stress
- High welding speed, which reduces heat input and improves productivity
- High-quality welds, which are free from defects, porosity, and oxidation.
- Ability to weld dissimilar metals, alloys, and refractory materials
- Low heat affected zone, which preserves the mechanical and metallurgical properties of the base metal

2.2.2. Disadvantages

- High initial cost of equipment and maintenance
- Need for a vacuum chamber and pumping system, which limits the size and shape of the workpiece
- Need for precise alignment and positioning of the workpiece and the electron beam
- Potential hazards from high voltage, radiation, and X-rays

2.2.3. Applications

EBW is a versatile and precise technique that can be used for various applications across different industries. Some of the industries that use EBW are aerospace and defense, nuclear and power generation, automotive and transportation, medical and dental, and electronics. EBW offers high-quality, high-speed, and low-distortion welding for a wide range of materials and thicknesses.

A brief comparison of laser beam welding and electron beam welding processes is given in Table 2.1 [5,7].

2.3. Friction Stir Welding

Friction stir welding (FSW) is a solid-state joining technique that uses a non-consumable rotating pin to heat and stir the contact surfaces between the workpieces. The FSW process does not melt the material but plasticizes it below the melting point. It requires a special machine capable of providing high rotational speed, axial force, and traverse speed [1,7-9]. Wear-resistant and high-temperature-resistant materials such as tool steels, high-speed steel (HSS), Ni-alloys, metal carbides, and ceramics are used to make the tool. The geometry and size of the tool are determined by the material type, thickness, and joint design. A more comprehensive examination of friction stir welding is provided in Chapter 8. Figure 2.3 shows the schematic diagram of the FSW process.

The primary steps of FSW are given below:

- The workpieces are clamped together and aligned along the joint line.
- A rotating tool with a shoulder and a pin is inserted into the workpieces until the shoulder touches the surface.
- It generates heat by friction and plasticizes the material around the pin as the tool moves along the joint line.
- The tool forges and stirs the plasticized material, resulting in the formation of a solid-state bond behind the tool.
- The tool is withdrawn from the workpiece, leaving a hole that can be filled or machined.

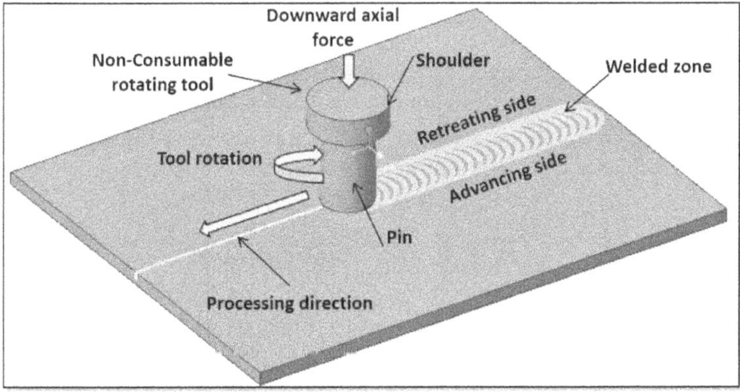

FIGURE 2.3 Schematic diagram of FSW

2.3.1. Processing parameters of FSW

A number of parameters affect the quality and performance (tensile strength, elongation, defects, hardness, etc.) of FSW welds [6]. As a result, it is critical to optimize the FSW settings in order to produce the desired weld characteristics. Some of the common parameters for FSW are given below [9].

Tool position: The depth and offset of the tool influence the heat generation, material flow, and joint quality.

Tool geometry: The shape and size of the tool shoulder and pin. It affects the stirring action, material flow, and forging effect of the tool.

Spindle RPM and travel speed: The rotational speed and linear speed of the tool determine the frictional heat input, stirring action, and welding time.

Tilt angle: The angle of the tool axis relative to the workpiece surface affects the material flow and pressure distribution. The angle at which the tool axis deviates from the vertical direction. It affects the pressure distribution, material flow, and asymmetry of the weld.

Joint design: The shape and size of the joint gap influence the material flow and defect formation.

Joint preparation: The cleanliness and flatness of the joint surfaces affect the frictional contact and bond strength.

The advantages and disadvantages of FSW are given in Table 2.2.

2.3.2. Applications

It is a newly invented welding technique that has found widespread use, primarily in the production of aluminum panels and components. It has many applications in various industries such as marine, fuel tanks, aircraft parts (fuselage panels, wing skins), boilers, military tanks, and more. FSW is increasingly used by the automotive industry to produce parts in large quantities, such as fuel tanks and light alloy wheels. It successfully welded polymers recently. FSW has also recently been able to join different metals together, such as aluminum and magnesium alloys.

2.4. Robotic welding

Robotic welding is a type of welding that uses robots to perform welding tasks. This process uses a robot arm controlled by specialized software to move the welding gun around the workpiece [10-13]. It can use different types of welding processes, such as gas metal arc welding (GMAW), resistance spot welding (RSW), or laser welding (LW). The robot moves the gun so precisely that it can make very intricate shapes or patterns with its welds. It is a highly advanced version of automated welding, in which machines conduct the welding, but welders still control and supervise the process.

There are some new advances in robotic welding technology that are shaping the future of manufacturing which are given below [11]:

- **Elimination of fixtures for increased efficiency**: Fabricators can use material handling robots to load and unload workpieces in the weld cell, reducing the need for manual intervention and saving time and costs.
- **Automated sensing for improved weld quality**: Robotic welders can use sensors to adjust their parameters according to the dimensions and distortions of each workpiece, ensuring consistent and accurate welds.
- **Augmented reality (AR) and virtual reality (VR) for training and simulation**: Welding students and professionals can use AR and VR technologies to practice and improve their welding skills in a realistic and safe environment, without wasting materials or risking injuries [10].
- **Seven-axis robots for enhanced flexibility**: Robotic welders with an additional axis in the lower arm can reach more complex positions and orientations, saving floor space and increasing productivity.
- **Internet of Things (IoT) for smart integration**: Robotic welders can talk to other devices and systems through the IoT. This lets them be programmed online or offline, have their parameters optimized, collect data, and keep an eye on the process.

2.4.1. Advantages and disadvantages

A robot welder can work faster and longer than a human welder, resulting in a reduction in cycle time and an increase in output. In addition, it reduces the risk of injury and improves safety by reducing exposure to hazards such as sparks, fumes, heat, and radiation. Moreover, robotic welding can reduce waste, errors, and rework costs by saving on materials, labor, and energy. Also, it is capable of producing high-quality, reliable welds that are uniform and accurate. In this way, robotic welding can improve manufacturing efficiency, profitability, and competitiveness.

On the other hand, it also has certain drawbacks. It has the potential to replace human labor, reducing their pay and employment options. It also needs expensive tools and infrastructure, which may be beyond reach for small and medium-sized firms. Furthermore, trained operators and technicians are required to program, maintain, and debug the robots. It is also not possible to detect defects or errors until the completeness after the weld is complete.

2.5. Hybrid welding

In hybrid welding, laser welding and arc welding are combined in the same weld pool to overcome the limitations of each method [1-3, 14]. A laser welding machine offers fast, precise, and deep welding, but it does not work with thick plates or different materials due to its high gap control requirements. On the other hand, arc welding is powerful against gaps and used for welding thick plates, but this method is time-consuming, superficial, and prone to distortion. Hybrid welding combines these two methods to get the benefits of faster welding, more depth, lower thermal distortion, harder and stronger welds, and improved seam quality [16-18]. Therefore, this welding technique is appropriate for welding thick plates, different materials, and intricate geometries. It is extensively employed in sectors including shipbuilding, car manufacturing, and rail vehicle construction.

There are various types of hybrid welding based on the type of arc welding mixed with laser welding. The list below includes the most typical types. Hybrid welding is discussed in greater detail in Chapter 10.

1. **Laser-MIG/MAG hybrid welding**: This hybrid welding process is the combination of the use of laser welding with metal inert gas (MIG) /metal active gas (MAG) welding, which uses a consumable wire electrode and an active gas shield. It is best suitable for welding of low-alloy steels, stainless steels, and carbon steels [17].
2. **Laser-TIG hybrid welding**: During this process, both laser welding and tungsten inert gas (TIG) welding take place simultaneously, which uses non-consumable tungsten electrodes and inert shielding gas. This method can weld both thin and thick materials as well as dissimilar metals and offers fine control and adaptability.
3. **Laser-plasma hybrid welding**: Similarly, this welding technique also involves the combination of two welding processes, laser welding, and plasma arc welding. The plasma arc bridges the gap between the workpieces, while the laser beam creates the keyhole effect, which requires a high-powered laser [16]. However, small and medium-sized enterprises cannot afford to purchase high-power lasers due to their high cost. A plasma-laser hybrid welding method is being developed in order to increase the productivity of existing methods (PAW and LBW) and to compensate for their drawbacks. It produces a highly economical welding technology. This process can enhance the penetration depth, stability, and efficiency of laser welding.

However, hybrid welding also has some disadvantages, such as:

- It requires a large number of process parameters to be checked and optimized for different welding sources and materials.
- It may cause residual stresses and distortion of the workpieces due to uneven heating and cooling.
- It may be prohibitively costly compared to conventional welding setups, especially when using high-power lasers.

2.6. Ultrasonic welding

Ultrasonic metal welding is a solid-state welding process that uses high-frequency ultrasonic acoustic vibrations to fuse materials by friction and pressure without melting them. It is commonly used for plastics and may also be used on certain types of metals, especially for joining dissimilar materials [19]. This process does not require any connective bolts, nails, soldering materials, or adhesives to bind the materials together. It can weld different types of materials, depending on their properties and compatibility. For plastics, ultrasonic welding can weld thermoplastics that have similar or different melting temperatures, such as polyethylene, polypropylene, nylon, ABS, and so on. For metals, ultrasonic welding can weld thin sheets or wires of metals that have similar or different electrical conductivity and resistance, such as copper, aluminum, nickel, gold, silver, and so on.

An ultrasonic metal welding machine works by following these steps [19-20]:

- **Preparation**: The fixture, also called the anvil or the nest, holds the two metal pieces that need to be joined. The pieces are arranged in the fixture before the welding starts.
- **Horn contact**: The horn, also called the sonotrode, is the part that delivers the vibration to the metal pieces. The horn touches the top piece and pushes it down toward the fixture.
- **Pressure application**: A press that is powered by air or electricity applies pressure to the horn and the pieces, keeping them in place.
- **Welding duration**: The horn vibrates up and down at a high speed and passes the vibration energy to the metal pieces in the fixture. The speed can range from 15 to 70 kilohertz (kHz) but is usually 20 kHz. The vibration creates heat energy at the point where the pieces touch each other, making them melt and join.
- **Hold time**: When the welding time is over, the vibration stops, and the system keeps applying pressure to the pieces, making them bond and form a strong weld.
- **Retraction**: After the weld has cooled and hardened, the pressure is removed, and the horn moves away, allowing the welded piece to be taken out of the fixture.

Although ultrasonic welding has numerous advantages, it is not the perfect welding process for all applications. There are some disadvantages of this process, which are explained in more detail below.

- It is limited to smaller parts that fit snugly together.
- It is generally more expensive than other welding methods.
- It requires clean surfaces and specialized equipment.
- It can cause distortion or weakening of the parts due to the vibrations.

2.7. Explosive Welding

Explosive welding is a solid-state metal joining process where two dissimilar metals are welded together by using a higher impact force which is generated by controlled detonation with an explosive charge.

The parts used in the explosive welding setup are listed below [21-22].

Target plate/base metal: It is the plate upon which cladding must be applied. The anvil is used to hold the target plate.

Flyer plate: The flyer plate is thinner than the target plate since it is constructed of cladding material. The flyer plate can be held at a fixed angle to the base plate, or it can be kept parallel to it.

Buffer plate: The flyer plate is protected from erosion caused by the explosion.

Explosive material: TNT, RDX, or any other type of explosive material can be used for explosive welding. Based on the size of the impact needed for welding, an explosive substance is chosen.

Stand-off distance: A stand-off distance is maintained between the base plate and flyer plate to allow for movement while the flyer plate moves over the base plate. The stand-off distance is approximately 0.5 to 1 times the flyer plate thickness.

Stand-off angle: The flyer plate is kept at an angle over the base plate for materials that cannot be welded by a constant stand-off distance. This angle is called a stand-off angle.

2.7.1. Working principle

A schematic diagram of explosive welding is shown in Figure 2.4. When the explosive detonates, it propels the flyer plate (cladding material) toward the base plate

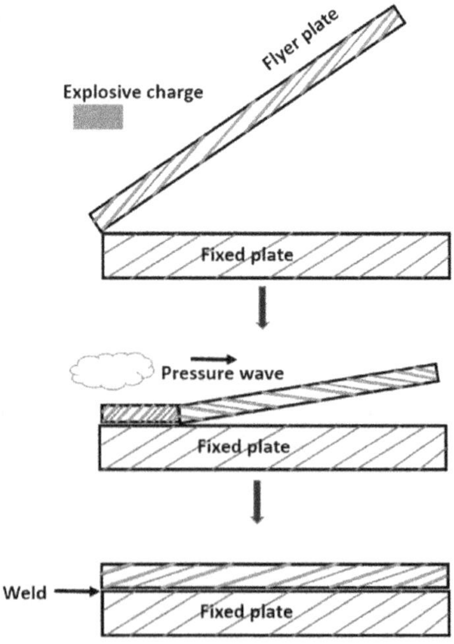

FIGURE 2.4 Schematic diagram of the explosive welding setup

(the target material) at high speed [23]. Due to this impact, plastic deformation and molecular diffusion take place between two plates, resulting in a joint formation. In this welding technique, the material is not melted and it does not use any filler material. It can also produce large-area welds with high bond strength and minimal distortion. It can be used for applications such as cladding, plugging, tubing, and so on. It allows to joining of dissimilar metals that cannot be welded by conventional means, such as carbon steel and stainless steel.

2.7.2. Disadvantages

- It requires extensive knowledge of explosives and special licensing before the procedure may be attempted safely. It may produce noise, blasts, and fumes that are harmful to the environment and human health. It has limited geometries that can be produced, such as plates, tubing, and tube sheets.

2.7.3. Applications

- It is used to clad carbon steel or aluminum plates with a thin layer of a harder or more corrosion-resistant material (e.g., stainless steel, nickel alloy, titanium, or zirconium).
- It is used to produce bimetallic pipes, tubes, and transition joints for cryogenic applications, such as liquefied natural gas (LNG) pipelines or heat exchangers.
- It is used to manufacture electrical transition joints between aluminum and copper for high-voltage applications, such as power transmission lines or electric locomotives [15].
- It is used to fabricate composite armor plates for military vehicles or structures, such as tanks or bunkers.
- It can bond large areas extremely quickly and produce clean welds.
- It does not require any external power source or shielding gas.

This explosive welding is used by various industries. For example, chemical and petrochemical industries use explosive welding to make corrosion-resistant clad plates for heat exchangers. Aerospace and defense industries make rocket nozzles, missile casings, aircraft structures, and so on. This welding method is also used by electrical and electronic industries to make electrical connectors, switches, circuit breakers, and other devices.

3. SUMMARY

The book chapter, "Various advanced welding methods: A brief overview," gives a summary of different ways of joining metal pieces with high-quality, low distortion, and high efficiency. The chapter describes in simple terms how each welding method works, what are its benefits and drawbacks, and where it can be used. It also discusses the main factors that influence the quality and strength of the welds. The techniques covered in the chapter are laser welding, electron beam welding, friction stir welding, ultrasonic welding, robotic welding, hybrid welding, and explosive welding. These methods have enabled industries to create new and innovative products by expanding the range of materials that can be welded. Through ongoing research and development, more advanced welding techniques will likely be available in the coming years.

REFERENCES

[1] J. Norrish, *Advanced welding processes technologies and process control*, Woodhead Publishing Limited Cambridge, England, 2006.

[2] Y. M. Zhang, Y. P. Yang, W. Zhang, and S. J. Na, Advanced welding manufacturing: A brief analysis and review of challenges and solutions. ASME, Journal of Manufacturing Science and Engineering 142 (11) (2020): 110816. https://doi.org/10.1115/1.4047947

[3] M. Chaturvedi, and S. A. Vendan, Ch-5, *Advanced welding techniques*. Springer, Singapore, 2021, 89–131. https://doi.org/10.1007/978-981-33-6621-3

[4] S. Thapliyal, In handbooks in advanced manufacturing. Advanced Welding and Deforming, (2021): 1–22. https://doi.org/10.1016/B978-0-12-822049-8.00001-3

[5] N. Ahmed, ed. *New developments in advanced welding*. Elsevier, 2005.

[6] A. Heidarzadeh, S. Mironov, R. Kaibyshev, G Çam, A. Simar, A. Gerlich, F. Khodabakhshi, A. Mostafaei, D.P Field, J. D. Robson, and A. Deschamps, "Friction stir welding/processing of metals and alloys: A comprehensive review on microstructural evolution. " Progress in Materials Science 117 (2021): 100752. https://doi.org/10.1016/j.pmatsci.2020.100752

[7] K. Mehta, "Advanced joining and welding techniques: An overview." Advanced Manufacturing Technologies: Modern Machining, Advanced Joining, Sustainable Manufacturing, (2017): 101–136. https://doi.org/10.1007/978-3-319-56099-1_5

[8] M. M. El-Sayed, A. Y. Shash, M. Abd-Rabou, and M. G. ElSherbiny, Welding and processing of metallic materials by using friction stir technique: A review. Journal of Advanced Joining Processes 3 (2021): 100059. https://doi.org/10.1016/j.jajp.2021.100059

[9] R. S. Mishra, and Z. Y. Ma. Friction stir welding and processing. Materials Science and Engineering: R: Reports 50 (2005): 1–78, https://doi.org/10.1016/j.mser.2005.07.001

[10] Y. Xu, and Z. Wang. Visual sensing technologies in robotic welding: Recent research developments and future interests. Sensors and Actuators A: Physical 320 (2021): 112551. https://doi.org/10.1016/j.sna.2021.112551

[11] P. Kah, M. Shrestha, E. Hiltunen, and J. Martikainen, Robotic arc welding sensors and programming in industrial applications. International Journal of Mechanical and Materials Engineering 10 (1) (2015): 1–16. https://doi.org/10.1186/s40712-015-0042-y

[12] T. S. Hong, M. Ghobakhloo, and W. Khaksar, Robotic welding technology. Comprehensive Materials Processing 6 (2014): 77–99.

[13] X. Fengjing, X. Yanling, Z. Huajun, and C. Shanben, Application of sensing technology in intelligent robotic arc welding: A review, Journal of Manufacturing Processes 79 (2022): 854–880, https://doi.org/10.1016/j.jmapro.2022.05.029

[14] F. O. Olsen, ed. *Hybrid laser-arc welding*. Elsevier, 2009.

[15] H. Krohn, and S. Singh, *Proc. Trends in welding of lightweight automotive and railroad vehicles*, Wels, Austria, 1997, 625–26.

[16] C. Emmelmann, M. Kirchhoff, and N. Petri. "Development of plasma-laser-hybrid welding process." Physics Procedia 12 (2011): 194–200.

[17] C. Bagger, and O. O. Flemming. "Review of laser hybrid welding." Journal of Laser Applications 17 (1) (2005): 2–14.

[18] M. Ono, Y. Shinbo, A. Yoshitake, and M. Ohmura, (2002). Development of laser-arc hybrid welding. NKK Technical Report-Japanese Edition, 70–74.

[19] S. K. Bhudolia, G. Gohel, K. F. Leong, and A. Islam. Advances in ultrasonic welding of thermoplastic composites: A review. Materials 13 (6) (2020): 1284. https://doi.org/10.3390/ma13061284

[20] M. P. Matheny, and K. F. Graff. *Ultrasonic welding of metals. Power ultrasonics.* Woodhead Publishing, 2015, 259–293.

[21] F. Findik, "Recent developments in explosive welding." Materials & Design 32 (3) (2011): 1081–1093. https://doi.org/10.1016/j.matdes.2010.10.017

[22] B. Crossland, and J. D. Williams. Explosive welding. Metallurgical Reviews 15 (1) (1970): 79–100.

[23] D. G. Brasher, and D. J. Butler. Explosive welding: Principles and potentials. Advanced Materials & Processes 147 (3) (1995): 37–39.

3 Heat Generation and Physics of the Arc Welding Process

Pavan Meena, Ramkishor Anant, and Nilesh Kumar Paraye

1. INTRODUCTION

Arc welding processes encompass intricate interplays of physical phenomena that result in heat generation, melting, and joining of materials. The fundamental principle underlying these processes is the utilization of an electric arc, which is a luminous discharge generated through the flow of current between two electrodes [1]. This arc generates intense heat, causing localized melting of the workpieces and creating a weld pool that eventually solidifies to form a strong joint.

The physics of arc welding processes involves several key aspects:

1.1. Electric arc generation

The welding arc is initiated by creating a potential difference between the electrode and the workpiece. The high voltage across the gap leads to ionization of the surrounding gas, forming a conductive path for the electric current to flow. This results in the establishment of the electric arc, a high-energy plasma discharge.

1.2. Plasma formation

The electric arc gives rise to a high-temperature plasma zone consisting of ionized gases, electrons, ions, and atoms. This plasma conducts electricity and emits intense light. The plasma temperature can reach several thousand degrees Celsius.

1.3. Heat generation

The arc's plasma zone generates significant heat due to the collisions between high-velocity electrons and gas molecules. These collisions transfer energy to the gas, leading to atomic excitation, which further elevates the temperature of the plasma.

 DOI: 10.1201/9781003435884-3

1.4. Excitation and Ionization

As ions and electrons recombine, they release energy in the form of light and heat.

1.5. Melting and Weld Pool Formation

The concentrated heat of the arc raises the temperature of the workpieces, causing localized melting. The molten material forms a weld pool, which serves as the medium for fusing the base metals together. The arc's energy must be appropriately controlled to prevent excessive penetration or insufficient fusion.

1.6. Heat Distribution and Transfer

The heat generated by the arc is distributed across the workpieces and surrounding environment. Heat flows through the weld pool, base metals, and even the electrode. The temperature gradient influences the size, depth, and integrity of the weld joint.

1.7. Shielding Gas

In some processes, a shielding gas is introduced to protect the welding area from atmospheric contamination. This gas also plays a role in stabilizing the arc, enhancing heat distribution, and controlling the weld pool's chemistry.

1.8. Electrode Contribution

The electrode material significantly influences the arc's behavior and heat generation. In processes like Gas Tungsten Arc Welding (GTAW), the non-consumable tungsten electrode maintains the arc while contributing to the heat input.

The physics of arc welding processes involves a delicate balance of electrical, thermal, and plasma dynamics. Factors like welding current, arc voltage, electrode material, shielding gas, and base metal properties all contribute to the final weld quality, penetration, and overall process efficiency. Understanding these underlying principles is essential for optimizing welding parameters, controlling heat input, and achieving desired welding outcomes.

Arc welding harnesses the energy concentration achieved through an arc plasma that forms between two metal electrodes. This focused energy is utilized to interact with the workpiece, composed of the metal components intended for welding. In this process, the workpiece undergoes partial melting, giving rise to a weld pool. The realm of arc welding delves into the intricate dynamics of heat transport, encompassing interactions between plasma, gas, liquid, and solid states of matter.

2. ARC INITIATION IN WELDING

The welding arc constitutes an electric discharge phenomenon, predominantly initiated by the flow of current from the cathode to the anode. For the current to traverse the gap between the electrode and the workpiece, the presence of a charged particle column is vital to ensure sufficient electrical conductivity. This column of charged particles originates from diverse mechanisms, including thermal emission, field emission, and secondary emission. The concentration of these charged particles within the gap significantly influences the electrical conductivity of the gaseous column. Arc welding operates on the fundamental concept of harnessing the arc's heat to liquefy the contacting surfaces of the metals intended for fusion. Initiating the welding arc marks the initial phase in forming a fusion weld. Within welding practices, two widely utilized techniques for commencing an electric arc are touch start and field start [2]. The field start approach is favored for automated welding tasks and in scenarios where electrode inclusion within the weld metal (WM) might occur during arc welding processes. On the other hand, the touch start method is employed across common manually controlled arc welding procedures. Within an electric arc, electrons are emitted from the cathode, a process instigated by either the electric field or thermo-ionic emission. Consequently, these electrons experience acceleration toward the anode, driven by the potential difference that exists between the workpiece and the electrode. While in motion, these high-velocity electrons traverse from the cathode to the anode, engaging with gaseous molecules in their path. This interaction prompts the dissociation of these molecules into charged particles—namely, electrons, and ions. Directed by their respective polarities, these charged particles subsequently journey toward both the electrode and the workpiece, synergistically constituting the overall welding current. Notably, the ion current constitutes a relatively small portion, approximately 1%, of the total electron current. This is due to ions having greater mass, resulting in slower movement. Ultimately, the electrons merge with the anode.

The arc gap existing between the electrode and the workpiece serves as a pure resistance load. The magnitude of heat produced within the welding arc hinges on two pivotal factors: the arc voltage and the welding current. These interdependent variables collectively determine the energy output of the arc, thereby exerting a profound influence on the welding process.

2.1. Touch start method

This approach involves initiating the welding process by bringing the electrode into direct contact with the workpiece and subsequently retracting it to create a narrow gap, typically ranging from approximately 2 to 4 millimeters, between the electrode and the workpiece. When the electrode makes contact with the workpiece, an electrical circuit is completed, causing a brief but intense surge of current to pass through the small contact area between the electrode and the workpiece. This surge leads to a momentary short-circuit scenario. The concentrated current flow within this confined contact region generates resistance-induced heating, resulting in the rapid production of excess heat.

This heightened heat generation at the interface where the electrode meets the workpiece triggers partial melting and, in some cases, even slight vaporization of the metal. Shortly after contact is made, usually within seconds (as depicted in

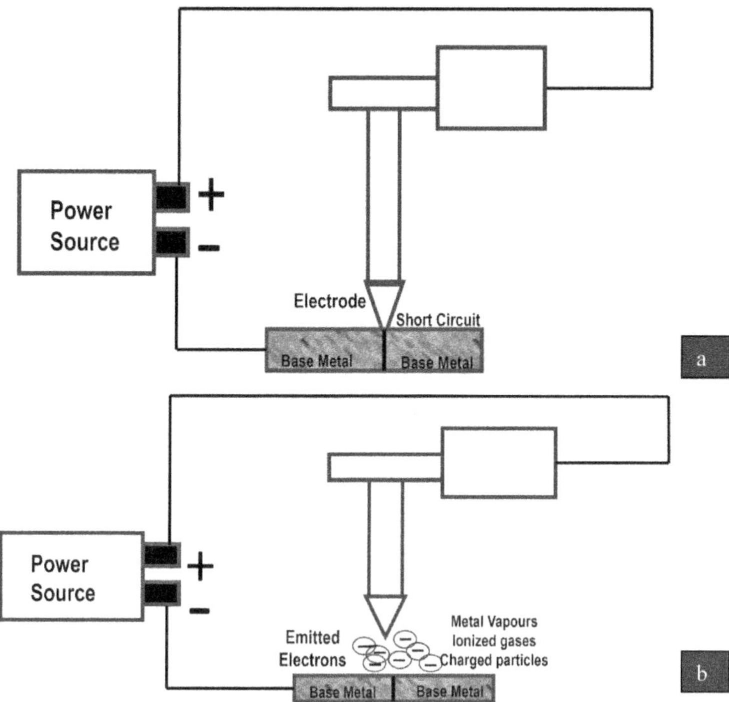

FIGURE 3.1 Diagram illustrating the touch start method arc initiation mechanism (a) when circuit is closed by briefly connecting the electrode with the workpiece, (b) when the electrodes are separated

Figure 3.1a, b), a sequence of events unfolds rapidly. The ionization process triggers the creation of metal vapor at the electrode-workpiece junction, which, in turn, yields charged particles such as ions and electrons. The thermal ionization of the electrode material and the comparatively lower ionization potential of metal vapors compared to atmospheric gases collectively result in the release of a small number of free electrons upon heating the electrode tip.

Introducing an adequate potential difference and delicately elevating the electrode a small distance away from the workpiece incites the flow of current across the gap between the electrode and the workpiece, now containing charged particles. This ensuing current flow precipitates a momentary arc or spark. To ensure consistent penetration, steady melting, and uniform fusion of the components, the electric arc's heat must be maintained and stabilized upon ignition. Nonetheless, this technique has two drawbacks: (a) a notable wear rate for the non-consumable electrode; and (b) the potential for mixing electrode metal with the weld metal. Both of these limitations primarily stem from the requirement of physically touching the workpiece with the electrode and the consequent thermal effects generated at the electrode-workpiece interface.

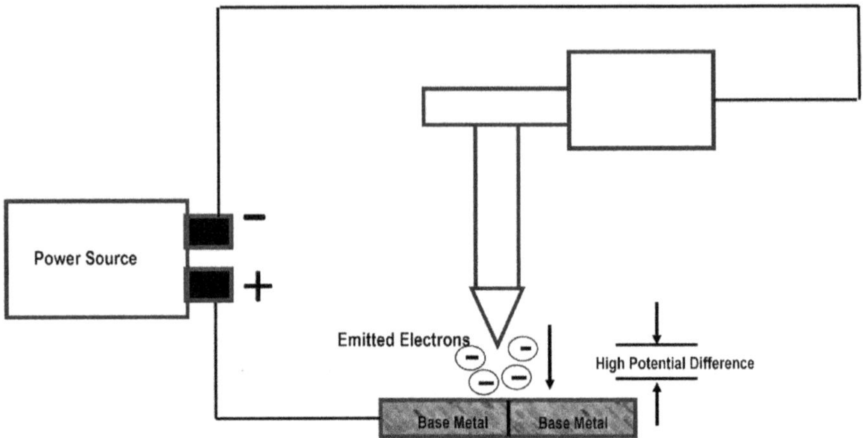

FIGURE 3.2 Illustration of field start technique for arc initiation

2.2. Field start method

To facilitate the emission of electrons from the cathode through electromagnetic field emission, a robust electric field is applied between the electrode and the workpiece, as depicted in this manner (refer to Figure 3.2). A substantial potential difference (on the order of 107 V) is employed to generate a potent electromagnetic field, either between the electrode and the workpiece in processes like GTAW or between the electrode and the nozzle, as observed in plasma arc welding (PAW) [2]. The presence of a magnetic field makes it easier to expel electrons from cathode spots. Once free electrons are introduced into the arc gap, the typical potential difference ranging from 10 to 50 V ensures the flow of charged particles required to maintain the continuity of the welding arc.

This arc initiation technique finds frequent application in mechanized arc welding methods such as plasma welding and gas tungsten arc welding. In these processes, direct contact between the electrode and the workpiece is avoided due to concerns related to (a) the formation of tungsten (W) inclusions and (b) the limited lifespan of the electrode. Alongside these factors, electrodes experience accelerated degradation due to contamination and thermal stress.

2.3. Scratch start method

Similar to touch start, the electrode is briefly scratched against the workpiece to create a short circuit, which triggers the arc. This method is used in some applications of GTAW and other processes.

2.4. Pilot arc

Some advanced welding processes use a pilot arc, which is a low-energy arc initiated separately from the main welding arc. The pilot arc ionizes the gas and preheats the electrode, making it easier to initiate the main welding arc.

2.5. Plasma arc torch

In plasma arc welding or cutting, a plasma arc torch is used to initiate the arc. The torch generates a high-temperature plasma stream that can pierce through the workpiece material, creating a path for the main welding arc.

2.6. Electron beam welding

In this high-energy welding process, an electron beam is used to initiate the arc. The focused electron beam provides the energy required to initiate the welding process.

3. EMISSION OF FREE ELECTRONS

The presence of liberated electrons and charged particles between the electrode and the workpiece plays a pivotal role in both initiating and sustaining the welding arc. This phenomenon is heavily influenced by two key parameters: the work function and the ionization potential of the material.

Work function: The work function quantifies a material's inclination to release electrons. It signifies the energy (expressed in eV or J) required to free a single electron from the surface of the material. The cathode metal's ability to emit electrons hinges on this particular value of work function.

Ionization potential: Another factor influencing a metal's electron-emissive capability is its ionization potential. This parameter gauges the energy (measured in eV) necessary to extract an electron from an atom, essentially representing the atom's grasp on its electrons. Notably, diverse metals possess varying ionization potential values. For instance, metals like Calcium (Ca), Potassium (K), and Sodium (Na) display remarkably low ionization potentials (ranging from 2.1 to 2.3 eV). In contrast, metals such as Aluminum (Al) and Iron (Fe) exhibit higher values of 4 and 4.5 eV, respectively.

The mechanisms responsible for the emission of free electrons during arc welding include:

a. Thermo-ionic emission

This phenomenon, known as thermo-ionic emission, involves the ejection of electrons due to heating. However, thermo-ionic emission generally occurs at temperatures that leads to metal melting [3]. Consequently, refractory materials with high melting points, like tungsten and carbon, exhibit a tendency for thermo-ionic electron emission.

b. Field emission

This method involves extracting free electrons from a metal's surface by inducing a strong electromagnetic field. A high potential difference, often on the order of 10^7 V/cm, is established between the workpiece and the electrode to facilitate field emission.

c. Secondary emission

Swift-moving electrons, journeying through the arc gap from the cathode to the anode, engage in collisions with gaseous molecules. These collisions incite the disintegration of these molecules into atoms and charged particles, encompassing both electrons and ions. This phenomenon, recognized as secondary emission, actively enhances the assembly of charged particles within the arc region. Together, these processes facilitate the existence of liberated electrons and charged particles, which are imperative for the commencement and sustained progression of the welding arc.

4. MAINTENANCE OF ARC IN WELDING

Once the welding arc is successfully initiated, it must be carefully maintained throughout the welding process to ensure consistent heat generation, proper melting, and effective material fusion. The stability of the arc directly impacts the quality of the weld joint. Here are some key considerations for maintaining the welding arc:

4.1. Arc length control

The distance between the electrode and the workpiece, known as the arc length, plays a significant role in maintaining the arc. A consistent arc length ensures stable heat input and prevents the arc from becoming too long (extinguishing) or too short (causing erratic melting). Welding operators often adjust the arc length by maintaining a steady electrode position and controlling the welding speed.

4.2. Welding current and voltage control

The welding current and voltage levels need to be controlled within specified ranges to maintain a stable arc. Too low a current can result in arc extinguishment, while excessive current can lead to excessive melting and penetration. Properly controlling these parameters ensures the desired energy input into the arc.

4.3. Electrode position and angle

Maintaining a consistent electrode position and angle is crucial for arc stability. An optimal angle and positioning help direct the heat and molten material appropriately, ensuring even melting and fusion along the weld joint.

4.4. Electrode consumption

In some welding processes like shielded metal arc welding (SMAW), the electrode is consumed as it melts into the weld pool. Proper electrode consumption control is necessary to maintain a stable arc. As the electrode is consumed, operators need to adjust the electrode position to maintain a consistent arc length.

4.5. Shielding gas and flux

For processes like gas metal arc welding (GMAW) and gas tungsten arc welding (GTAW), maintaining a continuous flow of shielding gas is essential. The shielding gas prevents atmospheric contamination of the arc zone, ensuring stable heat transfer and reducing the risk of defects.

4.6. Electrode and torch cooling

In some processes, like GTAW, excessive heat can cause electrode or torch damage. Proper cooling systems should be employed to prevent overheating and maintain consistent arc performance.

4.7. Operator skill

The skill of the welding operator is a critical factor in arc maintenance. Experienced operators can make real-time adjustments to various parameters, ensuring the arc remains stable even in challenging conditions.

4.8. Arc control technology

Some advanced welding machines offer arc control features that help maintain a stable arc even with variations in parameters like travel speed and joint geometry. These technologies can enhance the overall quality and efficiency of the welding process.

5. METHOD INVOLVING LOW IONIZATION POTENTIAL ELEMENTS

In this technique, the electrode's flux or coating is infused with elements characterized by low ionization potentials, including potassium, calcium, and sodium. The intention is for these elements to release free electrons when exposed to the conditions of arc welding. Notably, this electron liberation takes place even in the presence of a relatively modest potential difference between the electrode and the workpiece [4]. The introduction of these liberated electrons intensifies the electrical conductivity within the arc, thereby facilitating the continuous and stable sustenance of the welding arc. Refer to Figure 3.3 for an illustrative depiction.

6. APPROACH UTILIZING LOW POWER FACTOR

The efficiency of power usage in an electrical system is measured by its power factor, with higher values being more favorable. However, in the context of welding, a distinct approach is employed. In welding, the power factor signifies the ratio of the real power employed for welding (extracted from the power source) to the apparent power drawn through the welding circuit.

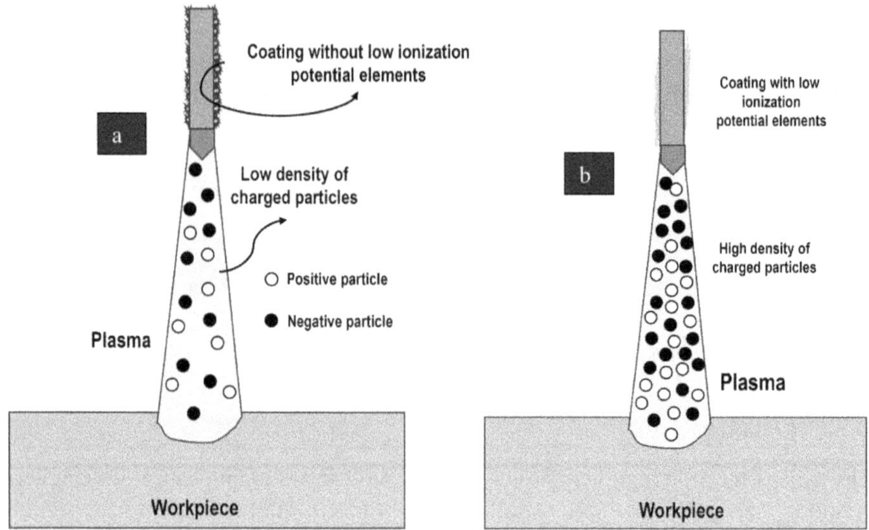

FIGURE 3.3 Illustrates the correlation between the density of charged particles within the welding arc under two scenarios: (a) in the absence of low ionization potential elements and (b) with the inclusion of low ionization potential elements [2]

Welding transformers typically function with a robust power factor that surpasses 0.9. Nonetheless, in the realm of AC welding, deliberately selecting a lower power factor proves advantageous as it enhances arc stability and guarantees efficient arc maintenance. In this technique, the alignment between current and voltage waveforms is purposefully altered to be out of phase, characterized by a deliberately low power factor (approximately 0.3). By employing this configuration, when the welding current hits zero, the electrode and workpiece are exposed to the entirety of the open circuit voltage, as indicated in Figure 3.4 at position "A". The substantial open circuit voltage encourages the liberation of additional free electrons, further enhancing the current of electrons already in motion. It is crucial to sustain an adequate concentration of charged particles between the electrode and the workpiece to uphold the continuity of the arc.

7. WELDING ARC CHARACTERISTICS

The welding arc characteristics refer to the variations in arc behavior, particularly in terms of voltage and current, as the welding process progresses. Understanding these characteristics is essential for controlling the welding parameters and achieving the desired weld quality. Moreover, the weld torch movement in case of numerical simulation [5] can be utilized to observe weld behavior with the groove and thermal cycle curve [6] in various heat-affected zones. The welding arc characteristic curve typically exhibits three distinct zones:

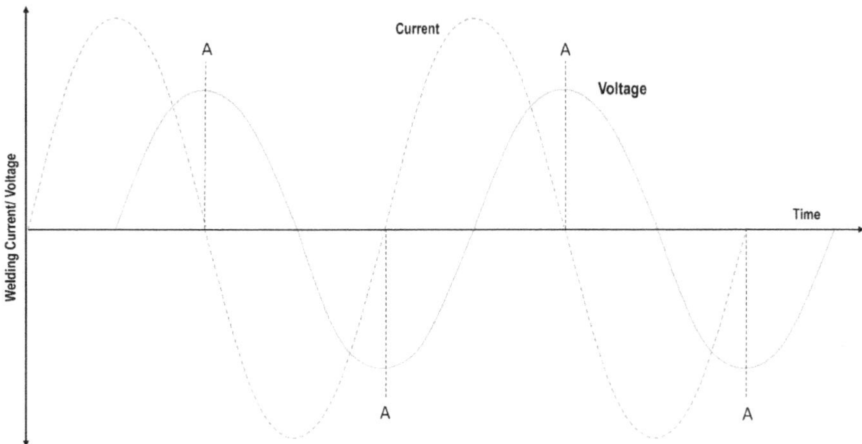

FIGURE 3.4 Current and voltage variations over time with a correct power

7.1. Drooping characteristic zone

This initial zone is characterized by a decrease in arc voltage as the welding current increases. At low welding currents, the arc is relatively thin, and increasing the current leads to higher temperatures and more charged particles in the plasma zone. This increase in charged particles results in improved electrical conductivity and reduced electrical resistance, causing the arc voltage to drop. The arc tends to stabilize within this zone.

7.2. Flat characteristic zone

Within this region, a rise in welding current exerts minimal influence on the arc voltage. As the welding current steadily escalates, the arc's diameter or cross-sectional area expands, causing heightened dissipation of heat from the arc's surface. Despite this augmented heat dissipation, the arc's temperature doesn't experience a substantial elevation. Consequently, the arc voltage remains fairly consistent across this spectrum of welding current.

7.3. Rising characteristic zone

Beyond the flat zone, further increases in welding current cause the arc voltage to rise. As the arc current increases, the arc diameter increases, leading to an elongated path for the current to flow. This extended path results in higher electrical resistance and greater heat losses from the arc surface. Consequently, the arc voltage starts to increase.

The welding arc characteristic curve provides valuable insights into the relationship between arc voltage and welding current, allowing welding operators to select

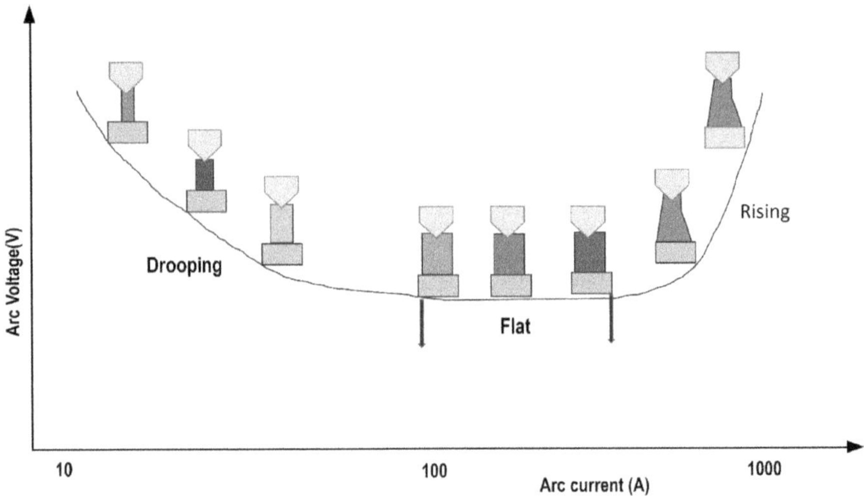

FIGURE 3.5 Welding arc characteristic curve [2]

appropriate parameters for achieving optimal weld penetration, fusion, and overall quality. The welding arc characteristic graph illustrates the fluctuation of arc voltage concerning welding current. This curve typically displays three distinct zones known as the drooping, flat, and rising characteristic zones, as depicted in Figure 3.5. During the initial stage, characterized by lower welding currents and a narrow arc, an increase in welding current leads to a rise in the arc's temperature. This rise in temperature leads to an increased concentration of charged particles within the plasma zone of the arc. This effect is a result of the thermal ionization of gases present in the arc gap and the thermo-ionic emission of electrons. As a result, the electrical conductivity of the arc zone intensifies, leading to a reduction in electrical resistance. This combination of factors leads to a decrease in arc voltage during the initial stages of current escalation, which falls within the drooping characteristic zone. This zone is associated with very low welding currents, and it is where the arc displays a propensity to stabilize.

This decrease in arc voltage with mounting current persists until a certain threshold of current is reached. Beyond this point, a surge in welding current leads to an expansion in the arc's diameter or cross-section, subsequently increasing the arc's surface area. The enlarged surface area facilitates a more noticeable dissipation of heat and charged particles from the arc's outer layer. As a result, the arc's temperature experiences modest growth as the current increases, leading to minor oscillations in arc voltage within the characteristic curve's level segment. Within this level segment, alterations in welding current have a limited impact on arc voltage, encompassing a range of values.

8. ARC TEMPERATURE CONSIDERATIONS

In addition to fundamental welding parameters like arc voltage and welding current, the temperature and its distribution across the arc region are notably influenced by the thermal properties (such as thermal conductivity and specific heat) of both the electrode and the shielding materials. Moreover, the ionization potential of the shielding gases within the arc vicinity exerts a substantial influence. In a broader context, three primary factors collaboratively shape arc temperature:

(a) **Increased arc power and ionization potential**: Amplifying the arc power and utilizing high ionization potential shielding gases both contribute to elevated arc temperatures. This is a consequence of a dual mechanism: greater heat generation resulting from higher arc power and ionization potential, and a reduction in heat loss from the arc surface due to the usage of shielding gases with lower thermal conductivity.
(b) **Influence of welding current and arc voltage**: Higher welding current and arc voltage lead to increased arc power, thereby affecting the arc's temperature distribution. The utilization of shielding gases with elevated ionization potential can also raise the arc voltage, subsequently boosting the arc's power.
(c) **Additional factors at play**: Several other variables come into play, including the thermal conductivity and diameter of the electrode, arc length, flow rate of shielding gases, and ionization potential of coatings, electrodes, or fluxes employed. These factors can either augment heat generation or mitigate heat loss, thus intricately influencing the dynamic temperature profile of the arc, as illustrated in Figure 3.6. Aspects such as higher thermal conductivity, larger electrode diameter, elongated arc length, and excessively high gas flow rates for shielding contribute to heat dissipation, resulting in a reduction of arc temperature. Conversely, the inclusion of elements with low ionization potential in coatings, fluxes, or electrodes can diminish heat generation, leading to a decrease in the arc's temperature.

Collectively, these variables that impact heat generation and transfer from the arc surface govern the temperature of the welding arc. This intricate interplay ultimately shapes vital welding characteristics, including penetration depth, melting rate, welding speed, overall productivity, and efficiency.

The thermal conductivity of numerous gases, such as helium (He), nitrogen (N), and argon (Ar), generally increases with rising temperatures. However, this pattern doesn't universally apply to all gases, notably helium. The interplay between the thermal conductivity of the base metal and the shielding gas significantly contributes to the temperature gradient within the arc region, as illustrated in Figure 3.7. When thermal conductivity is reduced, this gradient becomes more pronounced, resulting in a substantial decrease in arc temperature as the distance from the arc's center to its periphery increases, as depicted in Figure 3.8.

Inside the arc, the highest temperature is observed at the core, aligned with the electrode's axis. As the distance from the axis increases, the temperature rapidly diminishes. The anode and cathode regions typically experience lower temperatures compared to the plasma zone, due to the cooling effects of the electrode and the workpiece.

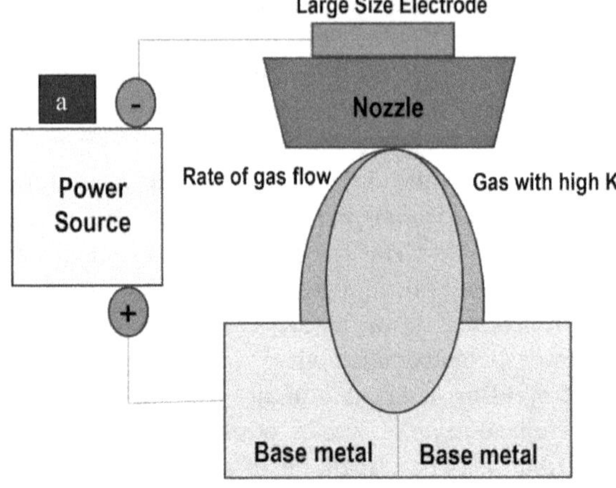

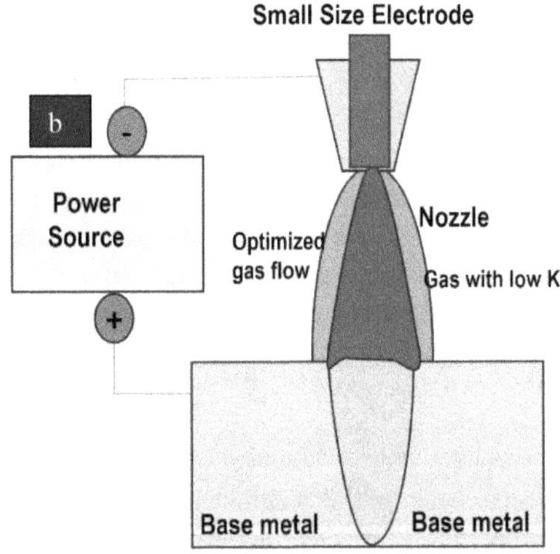

FIGURE 3.6 Depicts the impact of electrode size on arc size/temperature and bead geometry. The illustration shows two scenarios: (a) with a large electrode diameter and (b) with a small electrode diameter

The arc's temperature can vary extensively, ranging approximately from 4000 to 30,000 K, influenced by factors such as the welding process, welding current, arc voltage, shielding gas, and plasma gas. For instance, SMAW typically results in a maximum arc temperature of around 6000 K, whereas Tungsten Inert Gas (TIG) and Metal Inert Gas (MIG) welding arcs can exhibit temperatures ranging from 15,000 to 20,000 K [2].

9. HEAT TRANSFER IN WELDING

Heat generation in welding refers to the process by which thermal energy is produced and focused at the weld joint, ultimately leading to the fusion of materials. Welding

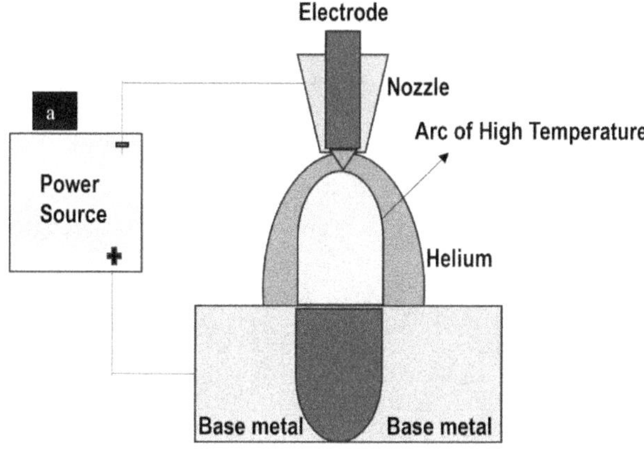

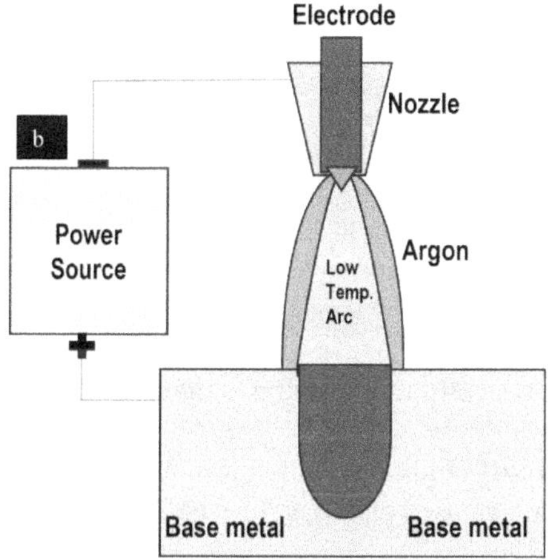

FIGURE 3.7 Shielding gas (a) helium and (b) argon on arc temperature and bead geometry

involves the union of two or more metal pieces by melting them at the joint and allowing them to solidify together. This fusion necessitates a considerable amount of heat.

The primary mechanisms through which heat generation occurs in welding are:

Arc heating: In processes such as gas metal arc welding (GMAW or MIG), gas tungsten arc welding (GTAW or TIG), and shielded metal arc welding (SMAW or stick welding), a welding electrode and the workpiece are brought together, initiating an electric arc. This electric arc emits high levels of heat, capable of exceeding the melting point of the base metal. The heat originates from the resistance encountered by the electrical current as it travels through the arc, causing the metals to heat up and ultimately melt.

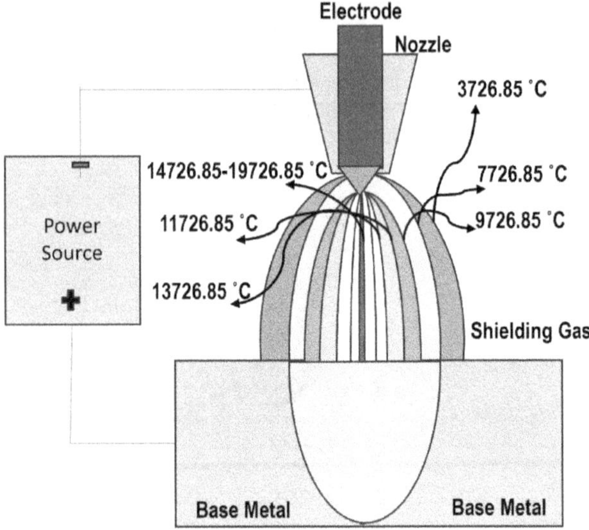

FIGURE 3.8 Temperature distribution in the arc [2]

Resistance heating: In processes such as Resistance Spot Welding (RSW) and Resistance Seam Welding (RSEW), heat is generated due to the resistance encountered by the electric current as it flows through the point of contact between the workpieces. This resistance leads to localized heat generation at the contact point, causing the metal to melt and fuse.

The extent of heat generation in welding is influenced by various factors:

a. **Welding current**: Higher welding currents typically lead to more heat production due to increased electrical resistance at the arc or contact point.
b. **Arc voltage**: Elevated arc voltages contribute to higher heat input, as they provide additional energy to sustain the arc.
c. **Travel speed**: Slower travel speeds allow more time for heat accumulation at the joint, resulting in increased heat input.
d. **Electrode material and diameter**: Different electrode materials and diameters impact heat transfer, consequently affecting heat generation.
e. **Properties of base metal**: The thermal conductivity, specific heat capacity, and melting point of the base metals influence their response to heat.

Effectively controlling heat generation is pivotal for achieving desired welding outcomes. Excessive heat can lead to distortion, overheating, and the deterioration of mechanical properties. Conversely, insufficient heat may lead to incomplete fusion or weak joints. Skilled welders manipulate welding parameters to strike the optimal equilibrium between heat generation and the integrity of the materials being welded.

Heat transfer in welding involves the complex interplay of conduction, convection, and radiation processes. Mathematical formulations are used to describe and predict the heat distribution and transfer within the welding process. The equations and models provide insights into temperature gradients, heat flow, and other critical aspects of the welding arc and weld pool. Here are some common mathematical formulations used in heat transfer analysis during welding:

9.1. Heat conduction equation

The basic equation governing heat conduction in welding is the heat conduction equation as shown in Eq. (1), which describes how heat flows through a solid material [3,20].

$$\frac{\partial}{\partial x}\left(k_x \frac{\partial T}{\partial x}\right)+\frac{\partial}{\partial y}\left(k_y \frac{\partial T}{\partial x}\right)+\frac{\partial}{\partial z}\left(k_z \frac{\partial T}{\partial x}\right)+Q=\rho C\frac{\partial T}{\partial t} \qquad \text{Eq. (1)}$$

9.2. Heat transfer by convection

In welding, convection plays a role when a fluid (often a shielding gas) flows over the weld pool. The heat transfer equation for convective heat transfer is given by Newton's law of cooling as shown in Eq. (2)

$$\left(Q_C\right)=h_C\,A\left(T-T_\infty\right) \qquad \text{Eq. (2)}$$

9.3. Radiative heat transfer

Radiation is also considered in some welding processes. The radiative heat transfer equation is given by the Stefan-Boltzmann as shown in Eq. (3)

$$(Q_R)=\sigma\,\varepsilon\,A\left(T^4-T_\infty^4\right) \qquad \text{Eq. (3)}$$

10. METAL TRANSFER PROCESSES

Metal transfer in the weld pool is a crucial aspect of welding processes, where molten metal is transferred from the welding electrode to the workpiece, contributing to the formation of the weld joint. The metal transfer mechanism varies based on the welding method being used, such as GMAW, GTAW, or shielded metal arc welding (SMAW).

a. **Globular transfer:** In this mode of transfer, large droplets of molten metal form at the end of the electrode and detach due to their own weight. The droplets fall into the weld pool, causing spatter and erratic bead formation. This transfer mode is typically used with higher welding currents and electrode diameters in GMAW.
b. **Spray transfer:** In this mode, small, finely atomized droplets of molten metal are propelled from the electrode to the workpiece by the force of the electric arc. Spray transfer is characterized by a steady stream of droplets and is commonly used in high-current, high-voltage GMAW applications.

c. **Short-circuiting transfer:** This mode involves the electrode momentarily touching the workpiece, creating a short circuit. The short circuit results in rapid melting of the electrode tip, causing a small amount of metal to transfer across the arc gap and into the weld pool. Short-circuiting transfer is commonly used in low-current GMAW and some GTAW processes.
d. **Pulsed transfer:** This mode combines elements of both globular and spray transfer. Pulsed transfer involves intermittent high-current pulses followed by lower-current periods, allowing controlled metal transfer while reducing spatter. It is often used in applications requiring precise control over heat input and weld quality.
e. **Hot start and cold start transfers:** These are variations of short-circuiting transfer used in SMAW. Hot start transfer involves initiating the arc with a higher current to preheat the electrode and the workpiece, while cold start transfer begins with a lower current to prevent excessive heat input. The choice of metal transfer mode depends on factors like welding current, electrode type, material being welded, joint configuration, and desired weld characteristics. The proper selection of metal transfer mode contributes to achieving consistent weld quality, controlled penetration, and minimal spatter.

In summary, metal transfer in the weld pool is a dynamic process influenced by the welding method and parameters. Understanding the different transfer modes and their effects helps welders make informed decisions to create strong, reliable welds.

11. HEAT TRANSFER THROUGH PULSED GAS METAL ARC WELDING (P-GMAW)

It involves the intricate movement of thermal energy within the welding process. P-GMAW, a variation of GMAW, utilizes pulsed welding current, creating alternating cycles of higher and lower energy input. Heat transfer in P-GMAW involves transient processes, characterized by alternating phases of high and low heat input. During high-current pulses, more heat is generated, leading to increased melting of the metal. Subsequently, low-current or background pulses facilitate cooling and solidification.

P-GMAW offers several advantages compared to continuous GMAW, including reduced heat input, enhanced penetration control, minimized distortion, and decreased spatter. Proper adjustment of pulse parameters can yield a controlled heat-affected zone and improved overall weld quality. To optimize heat transfer in P-GMAW, welders must comprehend the impact of pulse parameters on heat input, penetration depth, and weld bead geometry. A combination of experimentation and precise parameter settings can enable effective heat management and desired welding outcomes.

11.1. Pulsing frequency and duty cycle

The frequency at which welding current pulses occur, along with the duty cycle (proportion of high-current duration in the total cycle), directly affects the rate of heat input. Employing higher frequencies and shorter pulse durations can result in controlled heat input.

11.2. Peak current (I_P)

Maintaining an optimal peak current in spray transfer mode is crucial to avoid undesirable outcomes, such as fluctuations in arc pressure that could lead to the intake of air into the inert jacket, subsequently destabilizing the arc. Excessively high peak currents are particularly concerning, as they can induce significant axial force due to the magnetic field generated by the electrode's current. This phenomenon becomes pronounced when numerous drops of relatively smaller diameters form, as the resulting arc forces might fragment or propel these drops laterally.

To ensure the appropriate magnitude and duration of the peak current, several researchers have turned to power law relationships (expressed as $I_{pn}t_P$ = constant, with n = 2) [7]. This empirical approach aids in determining the suitable peak current amplitude and its duration. Furthermore, researchers have also identified the minimum pulse amplitude for mild steel filler wire, as evidenced in previous studies [8].

$$I_{p.min} = \frac{2.85 \times 10^4 \sigma^{0.5} d_e f^{0.245}}{I_{eff}^{0.465} t_p^{0.6}} \quad \text{Eq. (4)}$$

In the provided context, the σ represents the surface coefficient, measured in N/m. The term "d_e" stands for the diameter of the electrode, expressed in millimeters. "f" denotes the frequency of pulse recurrence, measured in pulses per second. The symbol "I_{eff}" represents the effective current, which is defined as follows:

$$I_{eff}^2 = \left\{k_p I_p^2 + \left(1 - k_p\right) I_b^2\right\}^{1/2} \quad \text{Eq. (5)}$$

wherein k_p is pulse duty cycle defined as t_p/t_{pul}, in A, t_p is pulse time, in seconds.

11.3. Base current (I_B)

The fundamental role of the base current is to sustain the welding arc during the intervals between pulses. Adjusting the base current to its minimum value permits control over heat distribution within the weld pool, exerting a significant influence on the resulting bead shape [7,9]. Extremely low levels of base current can yield a weld bead with pronounced height, potentially leading to insufficient fusion at the sidewalls. Consequently, the appropriate levels of base current can exhibit notable variation, contingent upon the material being welded. For instance, in the case of mild steel, this range typically spans 30–50A, while it hovers around 50A for austenitic stainless steel, and approximately 20A for aluminum alloys. However, the determination of base current levels and their corresponding durations hinges largely on the mean current applied [7].

11.4. Pulse frequency (f)

The pulse frequency holds notable influence over both the minimum required base current and the minimum electrode melting rate, particularly when aiming to maintain a consistent arc length. When sinusoidal pulse frequency increases while maintaining a given base current, it results in heightened mean and effective currents, as well as an accelerated rate of electrode melting. However, when altering the pulse frequency while keeping the mean current constant, the base current must remain adaptable, guided by the energy balance principle, in order to achieve a constant arc length.

Based on references from the available literature [7], empirical observations have highlighted a distinct behavior pattern. At lower pulse frequencies, the metal transfer mechanism is characterized by the emission of large drops or aggregates, leading to a viscous appearance of the weld pool and erratic arc behavior. In contrast, an increase in pulse frequency prompts the transfer of smaller axial droplets. The size of these droplets is determined by the balance between the forces causing detachment at the peak current and the surface tension's retaining force.

As a result, determining the optimal pulse frequency primarily relies on the average current and necessitates pre-selection based on specific conditions. The theoretical computation of frequency, as expounded in Eq. (5) [9], entails dividing the electrode melting rate by the mass of an individual drop.

$$\text{Theoretical frequency} = \frac{m_{pulse}}{V_{drop}\left(I_p\right)\rho_d} \quad \text{Eq. (6)}$$

In this context, "m_{pulse}" signifies the rate of electrode melting achieved through the application of pulsing current. "$V_{drop}(I_p)$" pertains to the volume of the drop at its peak current, and "ρ_d" represents the density of the drop. In the case of a square wave current, the estimation of the average melting rate necessitates a calculated amalgamation of the melting rate occurring at the peak current and the melting rate observed during the base current. This interrelation is presented in Eq. (6).

$$m_{pulse} = Dm(I_p) + (1–D)m(I_b) \quad \text{Eq. (7)}$$

Here, "D" signifies the load duty cycle, while "$m(I_p)$" and "$m(I_b)$" represent the melting rates at the peak and base currents, respectively. At a notably high pulse frequency, typically around 100 Hz, the transfer of metal occurs in the form of small axial droplets. This situation is conducive to a stable arc due to the increased arc pressure, resulting in a broader weld pool. On the other hand, employing a lower pulse frequency of around 25 Hz allows for the utilization of a reduced mean current, a favorable scenario for welding thin sheets [7,9]. Despite these observations, a comprehensive understanding of the precise influence of pulse frequency on the welding process and resultant weld characteristics remains incomplete.

11.5. Pulse duration (T_p)

The manipulation of pulse duration serves as a variable that can be adjusted to achieve optimal reduction in power for the GMAW process, all the while maintaining the spray transfer mode of metal transfer. Altering the shape of the pulse impacts the power input to the arc, with square waves, for instance, exhibiting a broader effective width compared to sine or triangular waves [7].

Pulse duration stands out as a pivotal operational parameter since it predominantly governs the formation of the "neck" and the elongation of the pendent drop during welding. In cases where the pulse duration is too brief, the elongated drop may revert to a spherical shape after the pulse, while excessive duration could lead to the detachment of multiple drops. Consequently, the selection of an appropriate peak duration is essential to ensure the detachment of either a single molten drop per pulse [10] or multiple droplets per pulse [11-12], depending on the desired outcome. The approach of transferring one drop per pulse is commonly adopted for joining thin sections, whereas the transfer of multiple droplets per pulse, coupled with controlled arc characteristics and precise metal transfer behavior, finds extensive application in the Pulse-GMAW process for welding various materials with varying section sizes.

In the context of aluminum welding, there exists an empirical relationship that establishes a connection between the peak and background conditions using a combination of exponential and Lorentzian functions. This relationship serves as a valuable tool for deducing the minimum necessary time at peak current, facilitating droplet detachment at a specific peak current magnitude [11], as outlined in Eq. (7). It's worth highlighting that the values of "t_p" and "t_b" within this equation should be expressed in milliseconds. Moreover, it's notable that this threshold slightly reduces with higher peak current values.

$$t_p = \left[\left(496.1 \times \left(1 - e^{-0.003 \times I_b t_b}\right)\right) + \frac{270.1}{\frac{\left(I_b t_b - 188.2\right)^2}{8423.5}} \right] / I_p \quad \text{Eq. (8)}$$

The determination of peak duration was additionally explored by [9], employing the notion that the pulsation period (t_p) described by Eq. 6 should be such that it facilitates the transformation of a cylindrical shape into a spherical shape with a diameter equal to the wire diameter (d_e). This transformation can be approximated as the peak duration required for the welding process.

$$T = \frac{240 V_{drop}}{\pi d_e^2 W_f} \quad \text{Eq. (9)}$$

Where T = $t_p + t_b$, V_{drop} is the droplet volume, $\left(\pi D_d^3\right)/6$ (mm^3), D_d the droplet diameter (mm) and W_f is the wire feed speed (m/min).

11.6. Droplet Detachment Mechanism

In P-GMAW, the process of detaching a droplet can be segmented into four distinct stages: heating, drop growth, necking, and detachment. These stages are defined by specific time intervals, with current amplitudes potentially varying during each stage. The duration of the pulse holds significant influence when a continuous spray of droplets is transferred under the effect of peak current.

The time required for both droplet formation and subsequent detachment is inversely proportional to the amplitude of the peak current. Interestingly, this phenomenon is independent of the pulse duration. As the necking stage initiates, a defined period is needed for the droplet to detach. This duration primarily depends on the peak current and the wire diameter, regardless of the current level during detachment. The necking process, involving the plastic deformation of the heated electrode, is driven by the Lorentz force, with the filler wire melting within the necked region. Detachment of the droplet is facilitated through the vaporization of molten metal at the neck, induced by resistance heating. The speed of droplet detachment is influenced by how efficiently the fused metal is compressed into a droplet shape.

Thus, it's widely recognized that the metal transfer characteristics and the thermal behavior of weld deposits in P-GMAW are intricately governed by the pulse parameters. These parameters interact during the welding process and collectively dictate the characteristics of the final weld deposit [11].

11.7. Thermal behavior of weld

The rate of wire burn-off or melting rate in pulsed gas metal arc welding is formulated as per references [13-15] as follows:

$$V_{w(pc)} = \int_{0}^{t_{pul}} V_w t_{pul} dt_{pul} \qquad \text{Eq. (10)}$$

In the context of a square pulsed current waveform:

$$V_{w(pc)} = \left(V_w t_p + V_{wb} t_b\right) t_{pul}^{-1} \qquad \text{Eq. (11)}$$

In this equation, $V_w(pc)$ represents the wire burn-off rate, while V_{wp} and V_{wb} denote the wire burn-off rates during the peak time (t_p) and the base time (t_b) respectively. The expressions for V_{wp} and V_{wb} are as follows:

$$V_{wp} = A.I_p + B.E_w.I_p^2 \text{ Eq. (12)}$$

$$V_{wb} = A.I_b + B.E_w.I_b^2 \text{ Eq. (13)}$$

Therefore, Eq. (11) may be rewritten as

$$V_{w(pc)} = \left[\frac{\left(A.I_p + B.E_w.I_p^2\right)t_p + \left(A.I_b + B.E_w.I_b^2\right)t_b}{t_{pul}}\right] \qquad \text{Eq. (14)}$$

Eq. (14) may be rewritten as

$$V_{w(pc)} = \left[\frac{A\left(I_p t_p + I_b t_b\right) + BE_w \left(I_p^2 t_p + I_b^2 t_b\right)}{t_{pul}} \right] \quad \text{Eq. (15)}$$

Using $I_m = [(I_p t_p + I_b t_b)/t]$ in Eq. (15)

$$V_{w(pc)} = AI_m + BE_w \left(I_p^2 t_p + I_b^2 t_b\right) t_{pul}^{-1} \quad \text{Eq. (16)}$$

As $I_p^2 t_p >> I_b^2 t_b$ neglecting ohmic heating during base current period, Eq. (16) reduces to

$$V_{w(pc)} = AI_m + BE_w I_p^2 t_p t_{pul}^{-1} \quad \text{Eq. (17)}$$

Considering $f = 1/t_{pul}$ and $I_p^2 t_p = D_n$, Eq. (18) gives

$$V_{w(pc)} = AI_m + BE_w D_n f \quad \text{Eq. (18)}$$

By incorporating the equation Vd = AwVw/f into Eq. (18), a linear correlation has been derived [18] in the following manner:

$$\frac{f}{I_m} = \frac{AA_w}{V_d - BDA_w E_w} \quad \text{Eq. (19)}$$

Solving Eq. (19) for droplet volume V_d,

$$V_d = AA_w \left(\frac{I_m}{f}\right) + A_w E_w BD \quad \text{Eq. (20)}$$

Eq. (19) reveals that, when the wire feed rate remains constant, consistent droplet size can be achieved by establishing a fixed ratio (Im/f), where Im represents the peak current divided by the frequency. This principle is applicable within defined parameters of wire diameter, electrode extension, and detachment criteria.

Empirical observations in the context of a 1.2 mm diameter steel filler wire [17] have indicated the possibility of attaining a desired combination of peak current and its duration, resulting in the ejection of one droplet per pulse. In this context, the determination of the base current and its duration can be guided by a frequency of 50 Hz, corresponding to a mean current of 100A [2]. Employing this methodology not only facilitates efficient droplet transfer but also ensures a relatively consistent droplet volume across a range of mean current values. This relationship can be mathematically expressed as follows:

$$I_m/f = 2 \qquad \text{Eq. (21)}$$

By combining Eqs. (18) and (21)

$$V_{w(pc)} = AI_m + \frac{BE_w DI_m}{2} = I_m\left(A + \frac{BE_w D}{2}\right) = \bar{A}I_m \qquad \text{Eq. (22)}$$

Where, $\bar{A} = \left(A + \frac{BE_w D}{2}\right)$ is the modified burn-off factor

In the P-GMAW process, when factoring in ohmic heating throughout the duration of the base current, Eq. (22) simplifies to:

$$V_{w(pc)} = AI_m + BE_w I_{eff}^2 \qquad \text{Eq. (23)}$$

Where, $I_{eff}^2 = \left\{k_p I_p^2 + (1-k_p) I_b^2\right\}^{1/2}$ and k_p is pulse duty cycle defined as t_p/t_{pul}

I_{eff} may also be expressed as,

$$I_{eff}^2 = I_m^2 + k_p(1-k_p)I_e^2 \qquad \text{Eq. (24)}$$

By combining Eqs. (23) and (24),

$$V_{w(pc)} = AI_m + BE_w\left(I_m^2 + k_p(1-k_p)I_e^2\right) \qquad \text{Eq. (25)}$$

From Eq. (23) & Eq. (25)

$$V_{w(pc)} = V_{w(cc)equiv.} + BE_w k_p (1-k_p) I_e^2 \qquad \text{Eq. (26)}$$

Eq. (26) provides insight into several aspects of P-GMAW:

In pulsed current welding, the rate of wire burn-off exceeds that of continuous current welding, even when both methods are employed at the same equivalent welding current.

The burn-off rate is affected by the pulsed nature of welding, and the maximum burn-off rate is attainable when the duration of the peak current matches the duration of base current (k_p = 1/2). This alignment occurs when the peak current notably exceeds the base current. Observations also reveal that the melting rate under P-GMAW is practically higher than the rate predicted by the weighted sum of melting rates (for DC currents) at the peak and base currents. Additionally, it's noted that the dynamics of the power source, which encompass the rate of current pulse rise and fall, can significantly impact the wire melting rate at a given mean current.

P-GMAW has garnered attention for its advantages over traditional GMAW, particularly in enhancing weld characteristics and the properties of various ferrous and non-ferrous metals. Researchers [11-12, 18] have reported these benefits under diverse pulsation conditions. However, the widespread adoption of the P-GMAW process in fabrication is hindered by the complexity of selecting optimal pulse parameters. Variations in pulse parameters, including I_m, I_p, I_b, t_p, and f, exert substantial influence over weld characteristics. It's realized that altering one parameter has a cascading effect on the others, as they all adhere to the principle of energy balance. These variations significantly impact the thermal behavior and nature of metal transfer, consequently affecting the attributes of the resulting weldment. In this context, achieving precise control over weld characteristics within desired parameters can be attained by establishing correlations among pulse parameters. Further, automation in the welding industry with Artificial Intelligence, and Machine Learning applications with computational simulations [19-20] are now being used for optimized parameters for temperature distribution and residual stress inducement in the weldment.

REFERENCES

1. American Welding Society (1987) *Welding handbook*, 8th edn, vol. 1 and 2, USA.
2. Dwivedi DK, Dwivedi DK. Physics of welding Arc: Arc initiation and Arc maintenance. *Fundamentals of Metal Joining: Processes, Mechanism and Performance.* 2022:85–95.
3. Meena P, Anant R. On the interaction to thermal cycle curve and numerous theories of failure criteria for weld-induced residual stresses in AISI304 steel using element birth and death technique. *Journal of Materials Engineering and Performance.* 2023 Apr 10:1–9.
4. Nadkarni SV. *Modern arc welding technology.* Oxford & IBH Publishing Company; 1988.
5. Meena P, Anant R. Numerical Investigation of Weld Induced Residual Stress State Field for Improved Structural Integrity in AUSC Power Plant Applications. *In International Conference on Advances in Materials Processing: Challenges and Opportunities* 2022 Oct 17 (pp. 147–152). Singapore: Springer Nature Singapore.
6. Singh MP, Meena PK, Arora KS, Kumar R, Shukla DK. Experimental, analytical and numerical, investigation of peak temperature and cooling rate in butt joint weld of mild steel. *World Journal of Engineering.* 2023 Jan 11;20(1):29–42.
7. Palani PK, Murugan N. Selection of parameters of pulsed current gas metal arc welding. *Journal of Materials Processing Technology.* 2006 Feb 20;172(1):1–0.
8. Jacobsen N. Monopulse investigation of drop detachment in pulsed gas metal arc welding. *Journal of Physics D: Applied Physics.* 1992 May 14;25(5):783.
9. Praveen P, Yarlagadda, PKDV, Kang MJ. "Advancements in pulse gas metal arc welding". *Journal of Materials Processing Technology.* 2005;164–165:1113–1119.
10. Rajasekaran S, Kulkarni SD, Mallya UD, Chaturvedi RC. Droplet detachment and plate fusion characteristics in pulsed current gas metal arc welding. *Welding Journal-New York.* 1998 Jun 1;77:254-s.

11. Ghosh PK, Gupta PC. Use of pulse current MIG welding improves the weld characteristics of Al-Zn-Mg Alloy. *Indian Welding Journal*. 1996;29:24–32.
12. Goyal VK, Ghosh PK, Saini JS. Process-controlled microstructure and cast morphology of dendrite in pulsed-current gas-metal arc weld deposits of aluminum and Al-Mg alloy. *Metallurgical and Materials Transactions A*. 2007 Aug;38:1794–805.
13. Craig E. A unique mode of GMAW transfer. *Welding Journal*. 1987;66(9):51–5.
14. Craig EF. "Pulsed spray welding- A mode of weld metal transfer that should revolutionize the GMAW process". *Welding Journal*. 1987b;66(9):79.
15. Smati, Z. "Automatic pulsed MIG welding." (1986).
16. Lambentt, JA. "Assessment of the pulsed GMA technique for tube attachment welding". *Welding Journal*. 1989;68(2):35–43.
17. Quintino L, Allum CJ. Pulsed GMAW: interactions between process parameters. *II. Welding and Metal Fabrication*. 1984;52(3):126–9.
18. Ghosh PK, Dorn L, Goecke SF. Universality of correlationships among pulse parameters for different MIG welding power sources. *The International Journal for the Joining of Materials: JOM*. 2001;13:40–47.
19. Meena P, Farhan AM, Anant R, Saheb SH. Advancements in welding technologies. Automation in welding industry: Incorporating artificial intelligence. *Machine Learning and Other Technologies*. 2024 Feb 19:13–35.
20. Meena P, Kumar M, Singh M, Kumar Shukla D, Burela RG, Jhunjhunwala P, Gupta A, Pandey C. Numerical analysis of different SUS304 steel weld joint configurations using new prescribed temperature approach. *Transactions of the Indian Institute of Metals*. 2022 Jun;75(6):1649–68.

4 Defects Associated with Welding Techniques and Their Detection Methods

Ranjeet Kumar

1. INTRODUCTION

Welding is a fundamental industrial process that involves utilizing heat and/or pressure to weld two or more components of metal or thermoplastic materials together. It is a crucial method used in numerous industries, including construction, automotive, aerospace, and shipbuilding [1]. Welding is essential in the development of permanent and reliable structures or products. Welding entails localized heating and melting of the materials to be joined and the incorporation of filler material, if necessary, to form a strong joint. The heated and melted components cool and solidify, making a continuous joint able to endure mechanical stress and other environmental factors. Despite its ubiquitous use, welding is not without its challenges. Welding defects could occur during the welding process, impacting the strength and integrity of the weld. This chapter discusses the different types of defects associated with different welding techniques and their detection using different techniques. Welding defects can be caused by a variety of sources, including human mistakes, incorrect equipment setup, and environmental conditions. Understanding the causes is critical for defect prevention.

There are several common mistakes observed in the welding process such as

- Welding Technique: Welding defects can occur due to insufficient training or a lack of experience. The poor technique includes incorrect welding torch manipulation, an incorrect electrode angle, and inconsistent travel speed.
- Inadequate Cleaning: Defects can occur when rust, dirt, oil, or other impurities are not completely removed from the weld joint. These contaminants can weaken the weld and cause porosity, inclusions, or lack of fusion.
- Incorrect Welding Parameters: Incorrect welding parameter settings, such as heat input, voltage, current, or wire feed speed, can result in defects formation. Inadequate heat can result in a lack of fusion, whereas excessive heat can produce distortion or burn-through.

DOI: 10.1201/9781003435884-4

- Improper Joint Preparation: Welding defects can be caused by an inadequate joint design, an improper bevel angle, or an inadequate gap between the base metals. It is essential to follow correct joint preparation procedures for different welding techniques.
- Inadequate Material Preparation: Defects might occur due to inappropriate material selection, or improper preheating. To ensure a successful weld, different materials demand different preparation processes.

2. TYPES OF WELDING DEFECTS

Welding defects are imperfections in the welding process that sacrifice the integrity of weld joints. Welding defects may manifest in various ways, each with its implications for weld quality. This section discussed the different types of welding defects, their causes and possible remedies.

2.1. Porosity

The presence of small cavities or voids within the weld metal is referred to as porosity, which is a common welding defect, as shown in Figure 4.1 (a). It forms due to the entrapment of gas bubbles in the weld pool. These voids might weaken the weld and cause it to fail under load. This is mainly present in tungsten inert gas (TIG) welding, as shown in Figure 4.1 (b). Certain base metals, such as aluminum or stainless steel, are more prone to porosity due to their inherent characteristics.

TABLE 4.1
Causes and remedies of porosity welding defects

Causes	Remedies
1. The presence of moisture, oil, grease, or other impurities on the surface of the base metal or filler material can vaporize during the welding process and cause porosity. **2.** Inadequate shielding gas coverage or incorrect gas flow rates might result in insufficient weld pool protection and porosity. **3.** Inappropriate filler material with a high moisture content or improper handling might contribute to porosity formation. **4.** Fast travel speed does not allow gas to come out from the weld pool.	**1.** Ensure that the base metal has been adequately prepared and cleaned. It should be free from moisture, oil, grease, and other contaminants. Use suitable cleaning methods such as wire brushing, grinding, or chemical cleaning. **2.** Ensure that the shielding gas flow rate is set correctly for the welding and material being welded. There are no leaks in the gas supply system and the nozzle or gas cup is properly positioned to provide adequate coverage. **3.** Keep filler material dry and follow the manufacturer's instructions for proper handling and storage. **4.** Preheating the base metal and electrodes as per recommended temperature before welding.

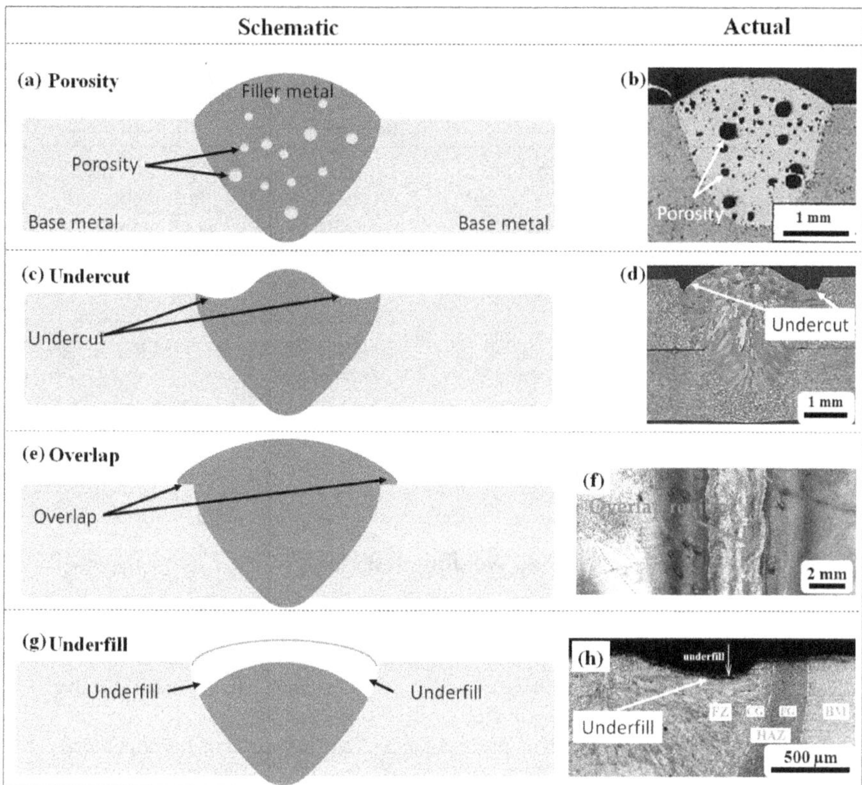

FIGURE 4.1 Illustrating schematic and actual micrographs of different welding defects (a, b) porosity [2], (c, d) undercut [3], (e, f) overlap [4], (g, h) underfill [5]

2.2. Undercut

Undercut is a welding defect that can be identified by the formation of a groove or depression at the base of the weld joint, along the weld toe or weld face, as shown in Figure 4.1 (c). It occurs when there is excessive melting of the base metal or not enough filler metal deposited in the joint. This results in a narrow cross-section, which means that there are notches or grooves along the weld, as can be seen in Figure 4.1 (d), which increases stress when the material is under fatigue, making it more prone to splitting or failure.

2.3. Overlap

It can be identified as an extra filler material that spreads around the weld bead and does not mix with the base material, as represented by Figure 4.1 (e, f). Hence, bonding is weak in this region.

TABLE 4.2
Causes and remedies of undercut welding defects

Causes	Remedies
1. Excessive heat input or rapid welding speed might cause the base metal to melt excessively, resulting in an undercut. **2.** Improper welding torch or electrode angle adjustment can also contribute to undercut creation. **3.** Inadequate beveling or joint fit-up can cause molten metal to flow into the joint gap, resulting in an undercut.	**1.** Maintaining an appropriate welding speed, managing heat input, and assuring proper welding torch or electrode angle control comprise all successful welding strategies. **2.** Ensure correct joint preparation, including suitable beveling, fit-up, and alignment.

TABLE 4.3
Causes and remedies of overlap welding defects

Causes	Remedies
1. It may occur if the welder uses poor welding procedures. This may include excessive weaving or zigzagging of the welding torch, which could end up in weld bead overlapping. **2.** It can be caused by improper selection of welding parameters such as current, voltage, or travel speed. **3.** Overlap defects can occur if the weld does not penetrate the entire thickness of the joint. **4.** It is also possible due to incorrect filler material.	**1.** Minimizes excessive weaving or zigzagging and maintains a stable welding speed. **2.** To promote better fusion, ensure adequate joint fit-up. **3.** Adjust the welding parameters to improve heat input to obtain better penetration if inadequate penetration is the cause of overlap defects. **4.** Consider employing preheating or using more suitable materials or filler metals.

2.4. Underfill

Underfill is a welding defect that arises when the weld bead does not fill the intended joint or groove adequately, as shown in Figure 4.1 (g, h), as a result, some of the parent material remains unfused. Even though they are minor, they act as potential stress raisers that compromise the welded joints' strength and integrity.

2.5. Incomplete fusion

It occurs when there is a lack of appropriate fusion between the filler metal and the base metal due to the pre-solidification of molten filler, resulting in gaps in the weld zone, as marked in Figure 4.2 (a, b). It can result in a weak joint and compromise the structural integrity of the weld.

TABLE 4.4
Causes and remedies of underfill welding defects

Causes	Remedies
1. Underfill can occur if the weld bead deposition is too small or the welder does not deposit enough weld metal. **2.** Underfill can be caused by poor joint preparation, such as wrong groove dimensions or inappropriate joint fit-up. Wide joint gaps cause underfill. **3.** Inconsistent or fast travel speed of the torch can lead to underfill. **4.** Underfill might occur from insufficient heat input during welding.	**1.** Optimize the welding current, voltage, and travel speed to provide sufficient heat input and adequate weld metal deposition. Increasing the heat input, such as raising the current or decreasing the travel speed, may help in proper fill and fusion. **2.** Maintain a steady weaving or oscillation pattern, offering appropriate weld bead overlap. **3.** Adjust the filler metal deposition using a larger diameter electrode or filler wire, or increasing the wire feed speed, can help achieve adequate fill. **4.** Ensure correct joint fit and alignment.

TABLE 4.5
Causes and remedies of incomplete fusion welding defects

Causes	Remedies
1. Inadequate heat input may not completely melt the base metal and achieve correct fusion, resulting in incomplete fusion. **2.** Incomplete fusion can be caused by improper weaving, excessive travel speed, or a wrong torch angle. **3.** An incorrect bevel angle may prevent proper fusion. **4.** Surface contaminants can also prevent fusion between weld metal and base metal. **5.** An incomplete fusion can occur when there is a gap or a poor fit between the joint surfaces. **6.** Certain materials with high thermal conductivity or those prone to surface oxide formation may be more difficult to weld with complete fusion.	**1.** Increasing the current, reducing the travel speed, or adjusting the voltage can help generate more heat, allowing for better fusion between the weld and base metal. **2.** To ensure adequate penetration and fusion throughout the weld, pay attention to the torch angle and distance from the joint. **3.** Properly clean and prepare the joint surfaces, ensuring the removal of any contaminants, scale, or oxides that can impede fusion. **4.** To enable effective fusion, ensure that the joint fit-up is correct, with proper alignment and minimal gaps. **5.** Choosing consumables with appropriate composition and melting characteristics will help to achieve better fusion. **6.** Preheating the base metal before welding can aid in improving fusion, especially in situations where incomplete fusion is a recurring issue.

TABLE 4.6
Causes and remedies of excess reinforcement welding defects

Causes	Remedies
1. Improper control of welding parameters such as high welding current and voltage could result in excessive welding metal deposition. 2. Improper electrode manipulation or weaving can result in excessive weld metal deposition. 3. Low travel speeds or an excessive filler wire feed rate can result in excessive weld metal deposition. 4. Welding in odd positions or angles can make managing the amount of weld metal deposited difficult, especially in places that are hard to reach or overhead welding. 5. Ineffective welder training, experience, or skills can all contribute to excessive reinforcement weld defects.	1. To minimize excessive metal deposition, maintain the optimal heat input and welding speed. 2. Control the feed rate of the filler wire to avoid excess reinforcement. 3. Remove excess weld metal with grinding tools or dressing processes to bring the reinforcement to the prescribed dimensions. 4. If the extra reinforcement defect is significant or cannot be effectively corrected through grinding or dressing, the weld may need to be completely removed and redo the welding.

2.6. Excess reinforcement

Excess reinforcement can be described as excess weld metal deposited in a joint. It is the inverse of underfilled welds, as can be seen in Figure 4.2 (c, d). With this defect, significant amounts of stress concentration might accumulate in the weld toes.

2.7. Incomplete penetration

Incomplete penetration is a welding defect that occurs when the gap between the base metals (root side) is not filled through the joint thickness, as shown in Figure 4.2 (e, f). This means that one side of the joint is not fused in the root which compromises the strength and integrity of the welded joint.

2.8. Excess penetration

Excess penetration is a welding defect that occurs when the weld metal penetrates the base metal too deeply, beyond the required or specified depth as shown in Figure 4.2 (g, h). This type of defect in pipe welding can cause fluid flow effects that might lead to erosion and/or corrosion issues.

2.9. Slag inclusion

Slag is a by-product of the welding process that consists of impurities, non-metallic elements, and fluxes. Slag inclusions are slag particles that get trapped in the weld during welding, as shown in Figure 4.3 (a, b). It compromises the final joint's integrity

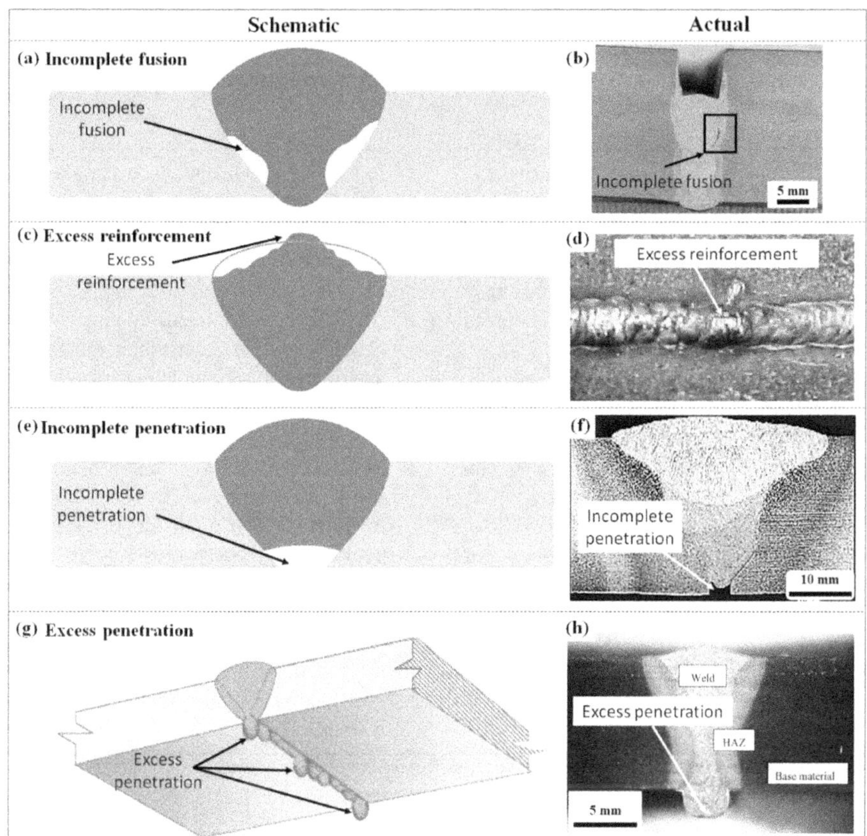

FIGURE 4.2 Illustrating schematic and actual micrographs of different welding defects (a, b) incomplete fusion [6], (c, d) excess reinforcement [7], (e, f) incomplete penetration [8], (g, h) excess penetration [9]

and makes the weld more prone to cracking or failure through them. This weld defect is common in flux-based welding processes such as stick, flux-cored, submerged arc welding, and brazing.

2.10. Spatter

Spatter refers to the undesirable scattering of molten metal droplets that occurs during the welding process, as shown in Figure 4.3 (c, d). It has the potential to produce rough and uneven surfaces and increase post-weld cleaning costs and time.

2.11. Arc strike

An arc strike is a welding defect that occurs when the welding arc accidentally touches a base metal surface other than the welding region, as shown in Figure 4.3 (e,

TABLE 4.7
Causes and remedies of incomplete penetration welding defects

Causes	Remedies
1. Incomplete penetration can occur when there is not enough heat generated during the welding process, preventing proper penetration. 2. Improper weaving, excessively fast travel speed, or an incorrect torch angle may all lead to incomplete penetration. 3. Incorrect groove dimensions, low bevel angle, or faulty joint fit-up can cause incomplete penetration. 4. Incomplete penetration is more frequent in joints with thick materials or complex joint designs.	1. Increasing the current, decreasing the travel speed, or altering the voltage can help to generate more heat for appropriate penetration. 2. Maintain a consistent weaving or oscillation pattern to ensure proper heat distribution. Adjust the torch angle and distance from the joint to provide adequate coverage and fusion throughout the joint. 3. Make sure to check for appropriate joint fit-up, alignment, and bevel angles. Clean the joint surfaces of any impurities that could obstruct penetration. 4. In cases where thick materials or intricate joint designs make complete penetration difficult, preheating the base metal elevates the temperature of the base metal, making it more receptive to the weld metal.

TABLE 4.8
Causes and remedies of excess penetration welding defects

Causes	Remedies
1. Excessive heat input can cause the weld pool to become extremely fluid, causing it to penetrate deeply into the base metal. 2. It might also happen when the welding speed is too slow. 3. Excessive molten metal penetration can be caused by holding the torch or electrode too close to the workpiece, excessive weaving or oscillation, or an incorrect torch angle. 4. A large gap or an incorrect bevel angle might cause the molten metal to flow excessively into the gap, resulting in deeper penetration.	1. Review and modify welding parameters (welding current, voltage, and travel speed) to control the depth of penetration. 2. To achieve the appropriate penetration depth, avoid excessive weaving or oscillation and control the angle of the welding torch or electrode. 3. To prevent excessive penetration, ensure a minimal gap or bevel angle. 4. Provide proper training and guidance to welding operators that can help to improve welding quality and reduce defects.

TABLE 4.9
Causes and remedies of slag inclusion welding defects

Causes	Remedies
1. Poor cleaning of weld bead during multipass welding. Previous pass slag can be trapped during subsequent welding passes. 2. Incorrect welding techniques, such as a wrong electrode angle, travel speed, or weaving speed, can trap slag in the weld. 3. Slag inclusions can occur as a result of improper shielding or incorrect selection of shielding gas or flux. Slag inclusions in the weld can also occur if the molten weld metal contains excessive slag. 4. Poor joint design, such as a low bevel angle, a lack of root space, or an incorrect fit-up, can result in slag traps.	1. Welders must be trained and educated on proper welding practices such as electrode angle, travel speed, and weaving motion to reduce the chance of slag inclusions. 2. Use the suitable shielding gas or flux to cover the weld region well. 3. After each welding pass, thoroughly clean the weld bead to remove slag/flux. Suitable cleaning processes, such as wire brushing or grinding, can be used. 4. The bevel angle can be meticulously chosen to avoid entrapment of slag/flux.

TABLE 4.10
Causes and remedies of spatter welding defects

Causes	Remedies
1. Excessive welding current or voltage can cause more spatter due to the high heat generated, causing metal droplets to detach from the electrode or filler wire. 2. A high wire feed speed may result in an unstable arc and excessive spatter. 3. Inadequate shielding gas flow can result in inadequate weld pool protection, resulting in increased oxidation and spatter. 4. Contaminants on the base metal surface or consumables (electrode or filler wire) might vaporize and cause a spatter during welding.	1. Adjust the welding current, voltage, and wire feed speed to the optimum levels for the welding procedure and materials. 2. Ensure the gas flow rate and shielding gas coverage around the weld. 3. Clean the base metal surface thoroughly to eliminate contaminants such as corrosion, oil, paint, or moisture. Furthermore, ensure that the consumables are clean and free of surface contamination. 4. Before welding, apply anti-spatter sprays, gels, or compounds on the base metal. These materials make a protective barrier to the base metal, hence spatter removal is easier after welding. 5. To reduce spatter, specialized welding torches nozzles or contact tips can be used.

TABLE 4.11
Causes and remedies of arc strike welding defects

Causes	Remedies
1. Inadequate welding skills or incorrect electrode position might increase the probability of arc strikes. 2. This might occur as a result of improper electrode angles or excessive electrode stick-out. 3. If the metal surface is unclean or contaminated or electrical power is unstable, striking an arc can be difficult, and the electrode may accidentally touch the metal. 4. Before striking the arc, the welder brings the electrode too close to the base metal. 5. The welder uses an electrode that is worn or broken, making it more prone to sticking to the metal.	1. Ensure that welders are using proper welding techniques, such as maintaining the correct electrode angle, appropriate electrode stick-out, and using undamaged electrodes. 2. To reduce the possibility of arc strikes, ensure that the welding equipment is properly grounded. 3. Clean the surface thoroughly prior to welding to remove any contaminants, debris, coatings, or materials that could cause an arc strike.

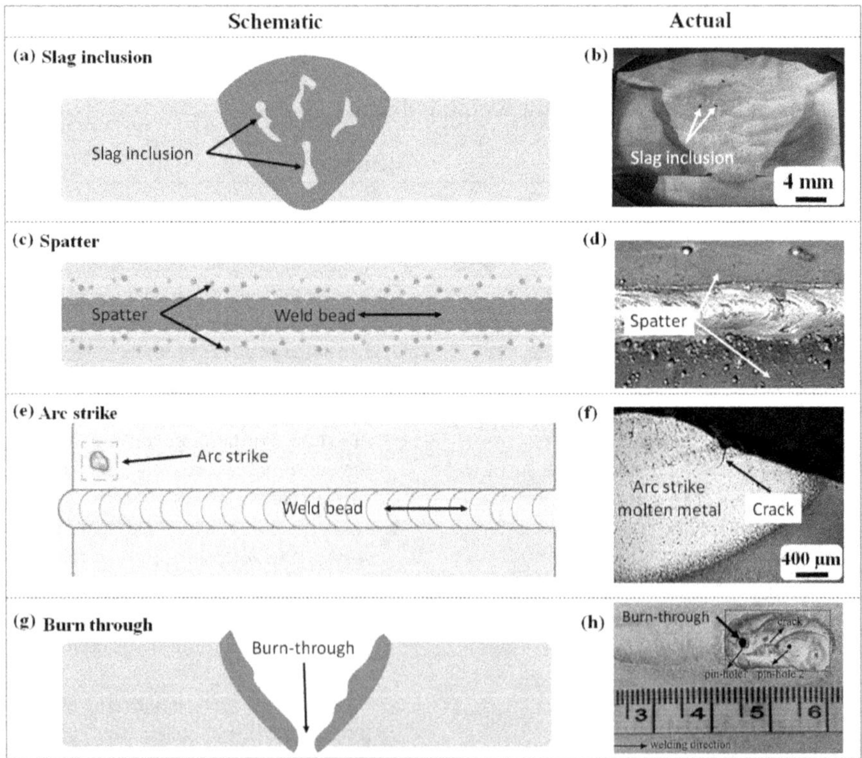

FIGURE 4.3 Illustrating schematic and actual micrographs of different welding defects (a, b) slag inclusion [10], (c, d) spatter, (e, f) arc strike [11], (g, h) burn-through [12]

TABLE 4.12
Causes and remedies of burn-through welding defects

Causes	Remedies
1. If the welding parameters, such as welding current or voltage, are set too high, the welding process can generate excessive heat, causing burn-through. **2.** Due to thin materials having less mass for dissipating heat, they are more prone to burn-through. Thin metal can quickly reach its melting point, resulting in burn-through. **3.** Inconsistent travel speed can result in concentrated heat in a single location, resulting in burn-through. **4.** A wider root gap is also the reason for burn-through.	**1.** Optimize the heat input (current, voltage) to avoid extreme temperatures that could result in burn-through. **2.** Controlling a steady and consistent travel speed to avoid excessive heat concentration in a single location. **3.** Ensure that the optimal joint root gap is maintained while welding. **4.** During welding, backer bars or heat sinks can help dissipate heat and support thin materials.

TABLE 4.13
Causes and remedies of distortion welding defects

Causes	Remedies
1. The metal expands due to the localized heat input, and contracts as it cools. This differential expansion and contraction can lead to distortion. It is more prevalent in thin-metal welding. **2.** Due to non-uniform heating and cooling, welding introduces residual stresses into the welded joint, leading to distortion. **3.** Inadequate alignment or significant gaps between joint surfaces might result in uneven heat distribution and distortion. **4.** The thermal expansion and contraction coefficients of base metal and weld metal are quite distinct. **5.** It also occurs when the number of weld passes is significant.	**1.** Make sure proper joint design, precise alignment, and minimal gaps between joint surfaces. This helps in the maintenance of uniform heat distribution during welding and minimizes the potential of distortion. **2.** During welding, use fixtures, clamps, or jigs to hold the components in the proper position. However, excessive constraint should be avoided since it could result in additional stress and subsequent cracking. **3.** Adopt a welding method that reduces temperature differences across the junction. **4.** Preheating the base metal can help reduce the temperature gradients and minimize distortion. Post-weld heat treatment, such as stress relief annealing, can alleviate residual stress and minimize distortion in some circumstances.

TABLE 4.14
Causes and remedies of misalignment welding defects

Causes	Remedies
1. This mostly occurs due to a lack of welder skills and not using the proper fixture.	**1.** Before welding, the joint surfaces must be carefully aligned.
2. This can be due to dimensional inaccuracies incorrect joint preparation, or improper part alignment.	**2.** To keep the components in place during welding, use fixtures, clamps, or tack welds.
3. Misalignment is more prevalent in rapid welding processes.	**3.** To avoid this type of defect, the welder should be skilled.

f). This can result in many problems including base metal deterioration, crack formation (see Figure 4.3 (f)), electrode and welding equipment damage, contamination of welds, localized hardness zones, and porosity.

2.12. Burn-through

Excessive heat causes the base metal to melt or burn-through, resulting in a hole or penetration in the base metal, as shown in Figure 4.3 (g, h). It mainly occurs when the heat input is too high or the base metal is too thin, resulting in burn-through.

2.13. Distortion

Warping or distortion is a modification in the geometry of the weld's surrounding metal, as shown in Figure 4.4 (a, b). Excessive heating around the weld joint is the primary cause of weld distortion. It is most common in thin metals and can be divided into four types: angular, longitudinal, fillet, and neutral axis.

2.14. Misalignment

A welding defect in which the components or pieces being welded are not correctly aligned results in an uneven or improper joint, as shown in Figure 4.4 (c, d). It can result in a variety of problems, such as weak welds, inadequate penetration, stress concentration and compromised structural integrity. This is more prone to fatigue failure, especially in pipe welding.

2.15. Welding cracks

Welding cracks typically can occur during and after welding. They are weld joint discontinuities that compromise the weld's strength and integrity. Welding cracks can occur in numerous forms including longitudinal, transverse, crater, hot and cold cracks, and root and edge cracks as shown in Figure 4.4 (e). Important ones are discussed in detail.

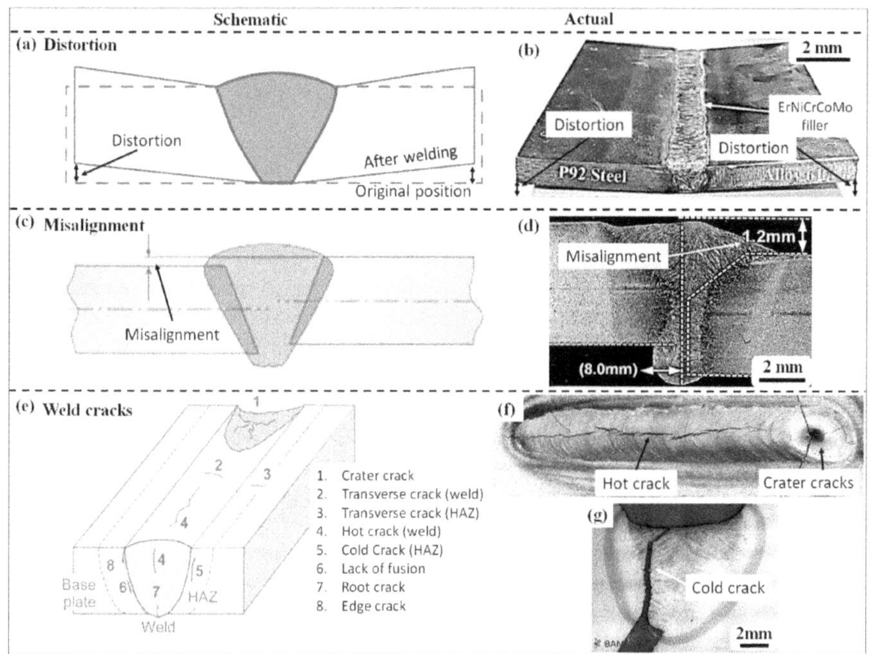

FIGURE 4.4 Illustrating schematic and actual micrographs of different welding defects (a, b) distortion [13], (c, d) misalignment [14], (e-g) weld cracks: hot crack [15], cold crack [16]

Note: Color images are essential for better understanding.

TABLE 4.15
Causes and remedies of crater cracks welding defects

Causes	Remedies
1. Because of an uneven distribution of thermal expansion and contraction, residual stress can build up in the weld crater, resulting in crater cracking. **2.** Crater cracks can form due to using some specific filler material with high cracking susceptibility, such as high-carbon steels.	**1.** The back-step technique, which involves welding the crater region first and then welding back over the previously solidified weld metal, can help mitigate the risk of crater cracks. **2.** Preheating the base metal prior to welding can help control the cooling rate and reduce residual stress. **3.** Filler materials with low crack susceptibility, such as low-carbon steels or suitable alloy compositions, can help in the prevention of crater cracks.

TABLE 4.16
Causes and remedies of hot cracks welding defects

Causes	Remedies
1. Some alloys contain low-melting-point elements or phases. These elements can segregate along grain boundaries during weld metal solidification, generating liquid films or layers, prone to cracking. **2.** Use of significant amount of sulphur and carbon. **3.** Weld metal/alloys with a narrow solidification temperature range solidify quickly, increasing the potential of localized melting and the formation of liquid films. **4.** Thermal and residual stresses can promote liquation cracking, such as those caused by rapid cooling or restrained solidification. These strains may be greater than the material's strength and cause fractures along the liquid films.	**1.** The choice of alloys with a larger solidification temperature range or compositions that minimize the presence of low-melting-point elements may help to mitigate liquation cracking susceptibility. **2.** Preheating the base metal before welding can help to reduce the cooling rate and the formation of liquid films, reducing the risk of cracking. **3.** Choosing filler materials in compatible compositions with the base metal can help to reduce the production of low-melting-point constituents and the susceptibility to liquation cracking. **4.** Post-weld heat treatment, such as stress relief annealing, may help decrease residual stress, minimizing the risk of liquation cracking.

TABLE 4.17
Causes and remedies of cold cracks welding defects

Causes	Remedies
1. During the welding process, hydrogen can be introduced into the weld. It could be caused by moisture, contaminants, or inadequate shielding gas. Hydrogen diffuses into weld metal and gets trapped in the solidified structure. **2.** Welding induces thermal expansion and contraction by introducing localized heating and rapid cooling. This causes residual stresses to form in the weld and HAZ. The combination of high residual stresses and hydrogen creates a conducive environment for cracking. **3.** Some materials are more susceptible to cold cracking such as high-strength steels, low-alloy steels, and certain stainless steels.	**1.** Preheating the base metal before welding can help decrease the cooling rate and reduce the potential of substantial residual stresses. Preheating also aids in the removal of moisture and the reduction of susceptibility to hydrogen-induced cracking. **2.** To avoid the introduction of moisture or contaminants that can contribute to increased hydrogen content, ensure proper storage and handling of welding consumables. **3.** Use low-hydrogen electrodes or filler wires designed to minimize hydrogen pickup during welding. **4.** Maintain adequate shielding gas coverage to protect the weld pool from moisture and air contamination. **5.** Post-weld heat treatment, such as stress relief annealing, may help release residual stress and reduce cracking risk. This method is especially advantageous for high-strength steel and other vulnerable materials.

2.15.1. Crater cracks

Crater cracks are specific types of welding defects that originate at the end of a weld bead and often take the shape of a small crack or fissure in the weld crater as shown in Figure 4.4 (e, f).

2.15.2. Hot cracks

Hot cracks, also known as solidification or liquation cracks or longitudinal crack, are welding defect that develops during the weld metal's solidification process. Hot cracks can form in the weld or heat-affected zone (HAZ), as shown in Figure 4.4 (e, g). They often appear shortly after welding is completed, while the weld material might be partially molten. It often occurs in alloys that undergo a eutectic reaction or have elements with low melting points.

2.15.3. Cold cracks

Cold cracking in welded joints, also known as hydrogen-induced cracking or delayed cracking, is a kind of defect that can occur after the joints have cooled. Cracks develop in the weld or HAZ as a consequence of the presence of hydrogen and substantial residual stress, as shown in Figure 4.4 (e, h).

TABLE 4.18
Causes and remedies of incomplete electrode selection

Causes	Remedies
1. Users may lack adequate knowledge of various electrode kinds and their uses. **2.** Choosing electrodes with inappropriate material properties for the workpiece or application. **3.** Using electrodes that are improper diameter or length for the job. **4.** Choosing electrodes with inappropriate coatings or alloys for the intended use. **5.** Ignoring the significance of electrode polarity during welding. **6.** Using electrodes below or beyond their recommended current and amperage ratings.	**1.** Proper training and education programs should be implemented to improve welders' knowledge for electrode selection depending on unique requirements. **2.** To choose electrodes from materials compatible with the workpiece and ensure optimal performance, conduct an in-depth study or consult professionals. **3.** To pick the suitable electrode size based on the current, material thickness, and welding position, consult manufacturer recommendations and charts. **4.** Consider the ambient conditions, needed mechanical qualities, and corrosion resistance before selecting electrodes with acceptable coatings or alloys. **5.** To achieve the required results and avoid complications such as improper penetration or spattering, adhere to the recommended polarity indicated for the electrode and welding technique. **6.** To prevent overheating, electrode degradation, and poor welding quality, strictly follow the manufacturer's current and amperage rating recommendations.

TABLE 4.19
Advantages and limitations of visual inspection method

Advantages	Limitations
1. Visual inspection requires less equipment, making it convenient and cost-effective for routine inspections and maintenance. 2. If defects are found, this allows for on-the-spot decision-making and quick measures, such as repairing the weld. 3. It is applicable for both surface defects and visually detectable discontinuities, making it a versatile method for identifying a wide range of welding defects in different materials. 4. Basic inspections require less training to execute. 5. It can be performed directly at the welding site, making it ideal for field or remote inspections.	1. Visual inspection is based on the inspector's interpretation and judgment. 2. It can only detect defects that are apparent to the human eye. 3. Certain sections or internal welds that are concealed or difficult to access may not be visually inspected. 4. Inspectors may miss or misinterpret defects due to different variables such as poor lighting, obstacles, or a lack of experience. 5. Visual inspection alone does not offer quantitative data on defects' size, severity, impact, or acceptance or rejection criteria.

2.16. Improper electrode selection

Improper electrode selection can result in inefficient performance and significant safety issues in welding applications. Here are some of the most prevalent causes and remedies for inadequate electrode selection.

3. DETECTION METHODS

Welding defects can have serious consequences for the strength and integrity of welded structures. Hence, detecting these defects is critical for ensuring weld quality. Many detection methods available nowadays can be chosen on a number of factors, including the type of defects to be examined, the material being welded, the required sensitivity, and the available resources and experience. Several testing methods are commonly used to ensure the comprehensive detection of defects and analysis. There are numerous methods for detecting welding defects, ranging from visual inspection to advanced non-destructive testing methods. Subsequent sections give the details of commonly used methods:

3.1. Visual inspection

The simplest and most basic approach to inspecting welds and their surrounding areas is to assess the quality and integrity of the welded joints. This inspection aims to find any visible indications of welding defects, such as cracks, surface porosity, incomplete fusion, incomplete penetration, undercutting, weld spatter,

TABLE 4.20
Advantages and limitations of dye penetration test

Advantages	Limitations
1. It is highly sensitive to surface cracks, defects, and even small discontinuities. **2.** It can be utilized for various materials like metals, polymers, ceramics and composites. **3.** It is a simple and quick process that can be performed on-site or in a laboratory. **4.** It is often inexpensive and requires less equipment and training.	**1.** It is unable to detect subsurface or hidden cracks or defects. **2.** It is not suitable for porous materials. **3.** The final results can vary depending on the skill of the individual. **4.** Contaminants, coatings, and surface irregularities can all interfere with test results, resulting in incorrect findings.

or improper bead shape. While this method is subjective and depends on the inspector's experience, it might be a good initial screening tool. It is critical to have adequate lighting and access to different viewing angles for visual assessment. Inspectors with knowledge and expertise of welding standards and acceptance criteria may detect and classify welding defects based on size, location, and severity. Visual inspection is frequently used as the first line of defence to identify welding defects, and it can be supplemented with other non-destructive testing methods to provide a more comprehensive assessment of weld quality. There are some portable tools used for visual inspections such as welding gauges, magnifying glasses, borescope, and so on.

3.2. Dye Penetrant Test

The dye penetration test, also known as liquid penetrant inspection, is a non-destructive testing method used to detect surface defects, cracks, or discontinuities. It is widely utilized in manufacturing, aerospace, automotive, and construction industries.

Working procedure:
First, clean the surface of the material with solvents or cleaning solutions to remove any dirt, grease, or contaminants. A liquid penetrant, typically a colored dye, is applied to the material's surface. It is permitted some time to penetrate surface defects by capillary action. Following the dwell time, the excess penetrant is carefully removed from the material's surface. Following that, a developer is applied to the surface, which is usually in the form of a white powder or a visible liquid. The developer helps in drawing out the penetrant from any discontinuities, making them more detectable. The typical process steps are shown in Figure 4.5 (a). As the developer draws out the penetrant, it leaves visible indicators on the surface, revealing any defects or cracks, see Figure 4.5 (a). The inspector then analyses the surface under proper lighting to determine if they are acceptable or requires further examination or repair.

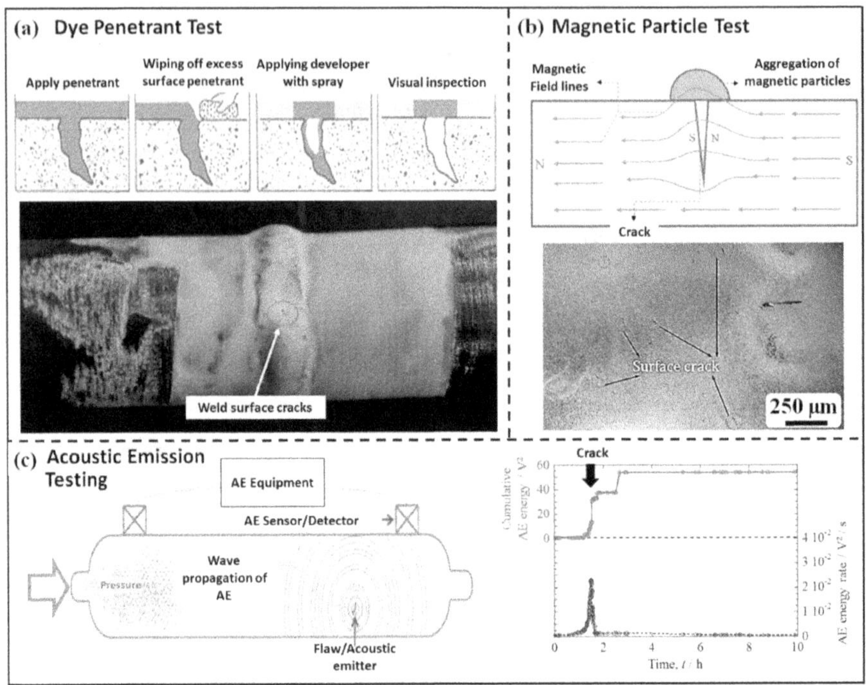

FIGURE 4.5 Illustrating schematic and actual pictures of (a) dye penetrant test [17] (b) magnetic particle testing [18], and (c) acoustic emission testing [19]

Note: Color images are essential for better understanding.

3.3. Magnetic particle testing

Magnetic particle testing is a method for detecting surface and near-surface defects in ferromagnetic materials. It is based on the magnetic flux leakage principle, see Figure 4.5 (b). Magnetic field lines travel through ferromagnetic materials when they are magnetized. If there is a defect or a discontinuity in the material, such as a crack or inclusion, the magnetic field lines will be distorted and "leak" out of the surface.

Working procedure:
An electromagnetic yoke is used to magnetize the component or structure to be inspected. The magnetic field within the material aligns the magnetic domains. Following magnetization, ferromagnetic particles, often iron or iron oxide, are applied on the surface. These particles might be dry or wet and can be suspended in oil or water. Magnetic particles are attracted and accumulated in locations where magnetic field lines are disturbed, implying the existence of defects. Under adequate lighting, the inspector interprets the indications based on their size, shape, and location to determine the nature and severity of the defects. The cracks can be seen in Figure 4.5 (b), under appropriate light.

TABLE 4.21
Advantages and limitations of magnetic particle testing

Advantages	Limitations
1. It is capable of detecting cracks, porosity, inclusions, and other surface defects that are open or near the surface. **2.** It can be utilized on ferromagnetic materials of various shapes (flat/curved) and sizes (small/large). **3.** It is much cheaper than radiographic or ultrasonic testing. **4.** It consists of portable equipment and can be easily transferred to the inspection site. **5.** It is relatively straightforward to perform, and the equipment and materials required are user-friendly.	**1.** It is limited to ferromagnetic materials such as iron, nickel, cobalt, and their alloys. **2.** It may not discover defects that are deeper into the material. **3.** The surface must be carefully cleaned to ensure that magnetic particles may adhere to it and that any indications are not hidden by contamination. **4.** It is more effective in detecting defects that are perpendicular to the magnetic field. **5.** It is difficult to detect very small defects because they may not generate an adequate magnetic field to attract magnetic particles. **6.** Parts having complicated geometries may provide difficulties in conducting a comprehensive magnetic particle inspection.

TABLE 4.22
Advantages and limitations of acoustic emission testing

Advantages	Limitations
1. It is capable of detecting and locating the occurrence of either instantaneous or progressive damage, such as crack initiation and propagation. **2.** It detects minor defects like microcracks and delamination that other NDT methods can overlook. **3.** It can be used on a variety of materials, such as metals, composites, concrete, ceramics, and others. **4.** It is a non-destructive approach that can inspect a broad region. It is useful for assessing large structures like bridges, pipelines, and storage tanks.	**1.** The detection range of the sensors limits the effectiveness of finding defects. **2.** The interpretation and analysis are complex and require skilled operators. **3.** Localization mistakes are possible, especially in structures with irregular geometries or heterogeneous materials. **4.** It is sensitive to environmental conditions such as temperature, humidity, and external vibrations. **5.** Large structures may require many sensors, increasing the complexity and cost of the inspection setup. **6.** Calibration of testing equipment is crucial to ensuring reliable and precise results.

TABLE 4.23
Advantages and limitations of radiography testing

Advantages	Limitations
1. It facilitates the detection of internal defects.	**1.** X-rays or gamma rays, which may be hazardous to human health.
2. It is highly sensitive to even small defects.	**2.** The radiation can easily travel through thin materials without generating an image, making inspection challenging.
3. It has the ability to examine a reasonably big area in a single exposure.	**3.** It requires the use of competent interpretation by experienced inspectors or technicians.
4. Radiographic images can be taken on film or stored digitally, providing a permanent record of the inspection results.	**4.** Due to limited access to the object or component, it may not be possible in some cases.
5. Radiographic testing can be used on a variety of materials, including metals, polymers, composites, ceramics, and others.	**5.** Material density and thickness, radiation scattering, and other geometric complications can all cause artefacts in the radiographs.
6. Radiography testing equipment is available in various sizes, including portable machines that enable on-site inspections.	**6.** It can be time-consuming and expensive when compared to other non-destructive testing methods.

3.4. Acoustic emission testing

It involves monitoring and analyzing the acoustic signals generated by a material or structure that is stressed or deformed. It is generally utilized in real-time to identify active defects such as cracks. The detection of transient elastic waves released by the material during the release of energy associated with changes in its internal structure is the basis of acoustic emission testing, as shown schematically in Figure 4.5 (c).

Working procedure:
The acoustic emission sensors, which are primarily piezoelectric transducers or microphones, are placed strategically on the surface of the material or structure being investigated, as shown in Figure 4.5 (c). The number and location of sensors are determined by the size and shape of the investigating area, as well as the specific testing requirements. Stress is applied (gradually/rapidly) to the material or structure. Mechanical, thermal, or a combination of the two types of stress could be present. The application of stress is supposed to cause deformation and probable defect activity inside the material. The sensors detect and transform the material's acoustic waves into electrical signals. These signals are subsequently amplified, filtered, and analyzed by an acoustic emission system, including signal acquisition and analysis hardware and data visualization and interpretation software. Based on amplitude, duration, rise time, and frequency content, identify and characterize the location,

magnitude, and type of defect. The shape of the cumulative AE energy curve is used to predict the occurrence of cracking, as shown in Figure 4.5 (c).

3.5. Radiography testing

Radiographic testing is capable of examining the surface and subsurface defects of welds using X-rays or gamma rays. This method generates a radiographic image on film or digital detectors that may be analyzed for defects such as porosity, lack of fusion, or incomplete penetration, as shown in Figure 4.6 (a). It inspects welds and offers extensive internal information, making it a crucial tool for quality control and safety assurance in many industries, including manufacturing, aerospace, oil & gas, automotive, and construction.

Working procedure:

A source of X-rays or gamma rays is placed on one side of the object being tested, and a detector or film is placed on the opposing side. The radiation goes through the object and is captured by the detector on the other side. The resulting image, known as a radiograph, displays any defects, discontinuities, or structural imperfections in the material, as shown in Figure 4.6 (a). Then after, skilled technicians or inspectors can interpret the radiographs and determine if the inspected component meets the required standards and specifications.

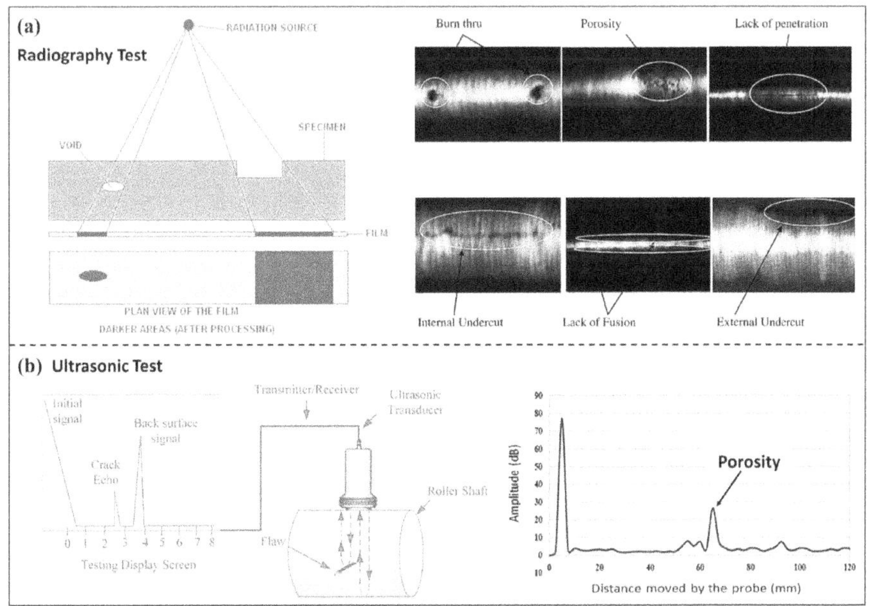

FIGURE 4.6 Illustrating schematic and actual pictures of (a) radiography test [20] and (b) ultrasonic test [18,21].

TABLE 4.24
Advantages and limitations of ultrasonic testing

Advantages	Limitations
1. It is highly sensitive to minute defects such as cracks, cavities, or inclusions.	**1.** Ultrasonic testing may be difficult or impractical in regions with restricted access or complex geometries.
2. It supports real-time imaging, allowing for immediate visualization and understanding of the investigated region.	**2.** It needs qualified and experienced specialists who can distinguish between typical sound wave patterns and indicators of defects.
3. It can be used on a variety of materials, including metals, composites, polymers, and ceramics.	**3.** Rough surfaces, surface coatings, and pollutants can interfere with ultrasonic wave transmission and reflection, thus hiding or masking defects.
4. It can be a more cost-effective inspection method than destructive testing or other complex inspection techniques.	**4.** Its efficiency decreases as the depth of the defect rises.
5. Ultrasonic testing is a non-hazardous inspection method that can inspect very thin material also.	**5.** Anisotropic materials and multilayer structures can make understanding ultrasonic signals challenging.

3.6. Ultrasonic testing

Ultrasonic testing detects and evaluates internal and external defects in materials by using high-frequency sound waves. It is frequently used to inspect welds, but it can also be used to inspect other components or structures.

Working procedure: A transducer is used in ultrasonic testing to generate ultrasonic waves and send them into the material being inspected. These waves propagate through the material until they reach a boundary, such as a defect or the material's rear surface. Some sound waves are reflected back to the transducer at the boundary, while others continue to travel through the material. The reflected waves are subsequently received by the transducer and transformed into electrical signals. The signals are analyzed to determine the time it takes ultrasonic waves to travel as well as the amplitude of the reflected signals. It is possible to identify the presence, location, and size of defects inside the material by investigating these characteristics, as shown in Figure 4.6 (b). Cracks, voids, lack of fusion, partial penetration, and inclusions can all be detected with ultrasonic testing. Ultrasonic testing involves several approaches, including:

Time-of-flight diffraction: This technology uses diffracted waves to identify and size defects precisely. It provides a more detailed inspection of welds, notably for size and characterizing cracks.

Pulse-echo testing: The most frequent technique, in which the transducer both generates and receives ultrasonic waves. It can be used to inspect materials for internal defects and material thickness.

Phased array ultrasonic testing: It uses many elements in the transducer to allow the beam to be electronically focussed and directed. This approach improves defect identification and characterization and is especially beneficial for examining complex geometries and welds.

It allows for the inspection of thick materials and provides a comprehensive evaluation of the volume.

3.7. Eddy Current Testing

Eddy current testing is a technique used to detect surface and near-surface defects in conductive materials. It operates on the basis of electromagnetic induction and the interaction of a magnetic field with electrical currents.

Working procedure: An alternating current is passed through a coil or probe to generate a changing magnetic field during eddy current testing. When the probe is brought close to a conductive material, the magnetic field causes tiny electrical currents in the material known as eddy currents, as shown in Figure 4.7 (a). The eddy currents generate magnetic fields that oppose the original magnetic field. The interaction of the

TABLE 4.25
Advantages and limitations of eddy current testing

Advantages	Limitations
1. It is sensitive to surface and near-surface defects such as cracks, pitting, corrosion, and material loss.	**1.** It works for shallow depths of penetration, often reaching a few millimeters or less.
2. It works relatively faster, making it useful for examining high-volume industrial applications.	**2.** Irregular or rough surfaces might produce signal interference or disperse eddy currents, lowering the reliability and precision of the inspection results.
3. It can be carried out without any physical contact between the probe and the material surface.	**3.** It is mainly designed for conductive materials.
4. It allows for customized probe configurations to meet specific examination requirements.	**4.** Eddy current signal interpretation is difficult and requires expert operators.
5. It provides immediate results, allowing for real-time examination.	**5.** It can be difficult to perform operations in complex geometries.
6. The technology is portable, making it ideal for on-site inspections and field applications.	
7. It can distinguish different types of defects based on their electrical conductivity, size, and position.	

TABLE 4.26
Advantages and limitations of infrared thermography method

Advantages	Limitations
1. Infrared thermography enables inspection without physical contact, which is especially beneficial when direct contact is impractical or risky. **2.** Infrared cameras can scan large regions quickly and produce real-time results. **3.** This allows for proactive maintenance and intervention before problems arise. **4.** Infrared thermography can be performed from a safe distance, lowering the risk to inspectors. **5.** It gives quantifiable data in thermal images, making comparison easy. **6.** It can detect heat loss and air leakage through cracks.	**1.** It can only detect surface defects (up to a few millimeters in depth). **2.** Materials with high reflective or low emissivity surfaces may not produce reliable results. **3.** Ambient temperature, humidity, air movement, distance, focus, and calibration all impact the accuracy and reliability of infrared thermography. **4.** Rapidly changing temperatures, moving objects, or time-dependent processes can all impact thermal images, making accurate interpretation challenging. **5.** Infrared thermography requires special tools, such as expensive infrared cameras.

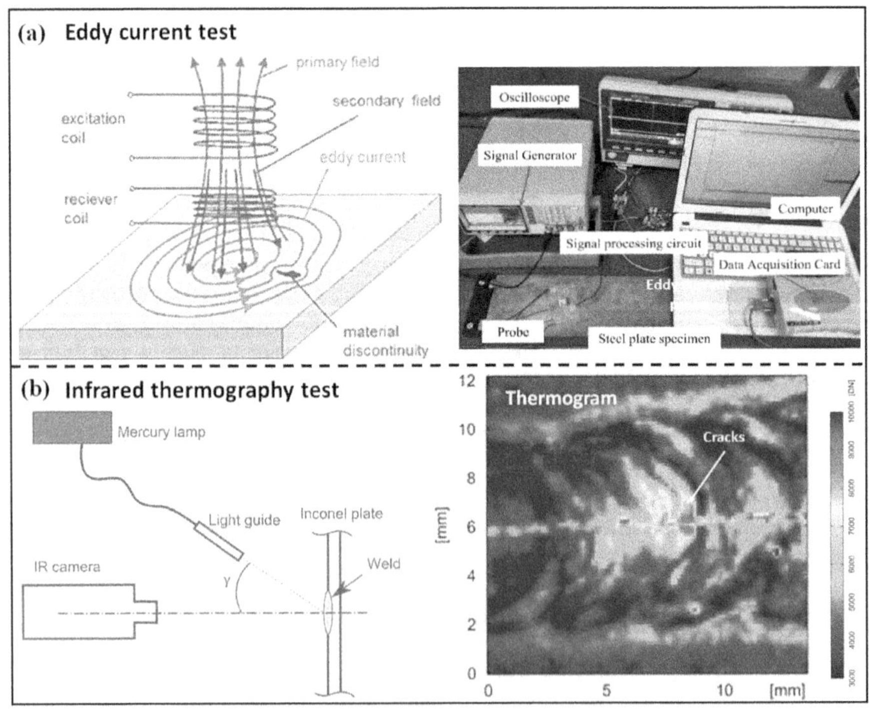

FIGURE 4.7 Illustrating schematic and actual pictures of (a) eddy current test [22] and (b) infrared thermography test [23]

Note: Color images are essential for better understanding.

two magnetic fields changes the electrical impedance of the coil or probe. An eddy current testing tool can identify and analyze changes in the material's electrical conductivity, magnetic permeability, or physical dimensions by measuring impedance changes, which can be indicative of surface or near-surface defects. Eddy current probes can be developed with a variety of designs, such as coils, flat coils, encircling coils, or customized probes for specific purposes. Cracks, corrosion, pitting, surface imperfections, and variations in material properties can all be detected via eddy current testing.

3.8. Infrared Thermography

Infrared thermography is a technique that measures the infrared radiation emitted by the surfaces of objects or structures to discover defects or anomalies. It is based on the fact that all objects with a temperature greater than absolute zero (-273.15 °C) produce infrared energy. Different materials have emissivity characteristics, meaning they emit and absorb thermal radiation differently. A defect in an object or structure frequently causes temperature changes or thermal patterns on its surface. These temperature variations can be visualized and analyzed using an infrared camera.

Working Procedure: During an infrared thermography inspection, the camera detects and measures the infrared radiation emitted by the surface, generating a thermal image or thermogram that illustrates the temperature distribution across the scanned region. This image is generally depicted in a false-color palette, where different colors indicate different temperature ranges, as shown in Figure 4.7 (b). Defects can be detected by analyzing the thermal image. Infrared thermography is a non-contact, non-destructive testing technique. It is widely utilized in a variety of industries, including building diagnostics, electrical inspections, mechanical inspections, and preventive maintenance.

4. SUMMARY

To summarize defects and detection methods, welding defects can substantially impact the strength, reliability, and safety of welded parts. The compromising of the strength and properties are associated with different defects such as porosity that weakens the weld by creating voids or cavities, while lack of fusion and lack of penetration result in insufficient bonding and limited load-carrying capacity. Undercut generates a groove at the weld joint's edge, reducing the weld's cross-sectional area. Spatter causes rough surfaces and possible contamination, lowering weld quality. Cracks in the weld or heat-affected zone dramatically decrease the strength and integrity of the joint. Inclusions, such as slag or foreign particles, weaken the weld and act as stress concentrators. Distortion can cause dimensional inaccuracies, misalignment, and stress concentrations, impacting the functionality of the welded component. Excessive weld reinforcement or underfill might affect joint strength and aesthetics. Welding defect detection methods are essential for identifying and analyzing defects in order to assure the quality and integrity of welded parts. They give essential details on the presence, type, and severity of welding defects, enabling proper repair operations and ensuring the quality and reliability of welded structures. The detection method chosen is determined by elements such as the type of defect, the criticality of the weld, cost considerations, and the available equipment and expertise.

REFERENCES

[1] P. Kah, J. Martikainen, Current trends in welding processes and materials: Improve in effectiveness, *Rev. Adv. Mater. Sci.* 30 (2012) 189–200.

[2] M. Pastor, H. Zhao, T. DebRoy, Continuous wave-Nd: yttrium–aluminum–garnet laser welding of AM60B magnesium alloy, *J. Laser Appl.* 12 (2000) 91–100. https://doi.org/10.2351/1.521922

[3] M. Idriss, F. Mirakhorli, A. Desrochers, A. Maslouhi, Overlap laser welding of 5052-H36 aluminum alloy: experimental investigation of process parameters and mechanical designs, *Int. J. Adv. Manuf. Technol.* 119 (2022) 7653–7667. https://doi.org/10.1007/s00170-022-08783-3

[4] S.H. Song, C.S. Lee, T.H. Lim, A. Amanov, I.S. Cho, Fatigue life improvement of weld beads with overlap defects using ultrasonic peening, *Materials (Basel)*. 16 (2023). https://doi.org/10.3390/ma16010463

[5] P.G. Riofrío, J. de Jesus, J.A.M. Ferreira, C. Capela, Influence of local properties on fatigue crack growth of laser butt welds in thin plates of high-strength low-alloy steel, *Appl. Sci.* 11 (2021). https://doi.org/10.3390/app11167346

[6] Z. Luo, Y. Li, H. Zhang, P. Wang, Experimental and simulation studies of micro-swing Arc welding process for X80M pipeline, *Metals (Basel)*. 13 (2023). https://doi.org/10.3390/met13071228

[7] S.M.M Alrewayeh, Defects resulting from fusion welding of metals and ways to prevent them, *Int. J. Eng. Res. Appl.* 12 (2022) 57–68. https://doi.org/10.9790/9622-12115768

[8] M. Boháčik, M. Mičian, R. Koňár, I. Hlavaty, Ultrasonic testing of Butt weld joint by TOFD technique, *Manuf. Technol.* 17 (2017) 7–8. https://doi.org/10.21062/ujep/x.2017/a/1213-2489/MT/17/6/842

[9] J. Górka, S. Stano, Microstructure and properties of hybrid laser arc welded joints (laser beam-MAG) in thermo-mechanical control processed S700MC steel, *Metals (Basel)*. 8 (2018). https://doi.org/10.3390/met8020132

[10] S.G. Parshin, A.M. Levchenko, A.S. Maystro, Metallurgical model of diffusible hydrogen and non-metallic slag inclusions in underwater wet welding of high-strength steel, *Metals (Basel)*. 10 (2020) 1–17. https://doi.org/10.3390/met10111498

[11] J. Chao, C. Peña, Effect analysis of an arc-strike-induced defect on the failure of a post-tensioned threadbar, *Case Stud. Eng. Fail. Anal.* 5–6 (2016) 1–9. https://doi.org/10.1016/j.csefa.2015.11.001

[12] H. Zhang, T. Han, Y. Wang, Study on the dynamic evolution behavior and failure mechanism of burn-through instability during In-Service welding by combining In-Situ observation and failure analysis, *Materials (Basel)*. 16 (2023). https://doi.org/10.3390/ma16031184

[13] A. Kumar, C. Pandey, Structural integrity assessment of Inconel 617/P92 steel dissimilar welds for different groove geometry, *Sci. Rep.* 13 (2023) 8061. https://doi.org/10.1038/s41598-023-35136-1

[14] L. Li, H. Xia, N. Ma, B. Pan, Y. Huang, S. Chang, Comparison of the welding deformation of mismatch and normal butt joints produced by laser-arc hybrid welding, *J. Manuf. Process.* 34 (2018) 678–687. https://doi.org/10.1016/j.jmapro.2018.07.015

[15] P. Yu, S. Kou, C.M. Lin, Solidification and liquation cracking in welds of high entropy CoCrFeNiCux Alloys, *Materials (Basel).* 16 (2023). https://doi.org/10.3390/ma16165621

[16] H.S. Steel, A.N. Look, Hydrogen-Assisted Cracking in GMA welding of high-strength structural steel—A new look into this issue at narrow Groove, *Metals (Basel).* 11 (2021). https://doi.org/10.3390/met11060904

[17] P. Mohyla, J. Hajnys, K. Sternadelová, L. Krejčí, M. Pagáč, K. Konečná, P. Krpec, Analysis of welded joint properties on an AISI316L stainless steel tube manufactured by SLM technology, *Materials (Basel).* 13 (2020) 1–14. https://doi.org/10.3390/ma13194362

[18] M. Jamil, A.M. Khan, H. Hegab, S. Sarfraz, N. Sharma, M. Mia, M.K. Gupta, G.L. Zhao, H. Moustabchir, C.I. Pruncu, Internal cracks and non-metallic inclusions as root causes of casting failure in sugar mill roller shafts, *Materials (Basel).* 12 (2019) 1–22. https://doi.org/10.3390/ma12152474

[19] T. Shiraiwa, M. Kawate, F. Briffod, T. Kasuya, M. Enoki, Evaluation of hydrogen-induced cracking in high-strength steel welded joints by acoustic emission technique, *Mater. Des.* 190 (2020) 108573. https://doi.org/10.1016/j.matdes.2020.108573

[20] B. Mhamed, S. Abid, F. Fnaiech, Weld defect detection using a modified anisotropic diffusion model, *EURASIP J. Adv. Signal Process.* 2012 (2012) 1–12. https://doi.org/10.1186/1687-6180-2012-46

[21] S.E. Florence, R.V. Samsingh, V. Babureddy, Artificial intelligence based defect classification for weld joints, *IOP Conf. Ser. Mater. Sci. Eng.* 402 (2018). https://doi.org/10.1088/1757-899X/402/1/012159

[22] T. Chen, H. Shi, Y. Dong, C. Lv, Z. Deng, X. Song, Design and performance research of a new dual-excitation uniform Eddy current Probe, *Sensors* 22 (2022). https://doi.org/10.3390/s22228850

[23] P. Broberg, Surface crack detection in welds using thermography, *NDT E Int.* 57 (2013) 69–73. https://doi.org/10.1016/j.ndteint.2013.03.008

5 Role of Filler Materials on Welding Performance

Nilesh Kumar Paraye, Ramkishor Anant, and Himanshu Vashishtha

1. INTRODUCTION

The evolution of various advanced steel and alloys tends to the adaptation in the engineering industry. The adaptation of these alloys in the engineering industry requires the process of joining and mainly welding from the industrial point of view. Nevertheless, these alloys, such as advanced strength steel, aluminium, nickel, and titanium, can be successfully welded with various welding techniques such as gas tungsten arc welding (GTAW), laser welding, shielded metal arc welding (SMAW), gas metal arc welding (GMAW), and so on. [1]. However, when joining these alloys, filler metal performs a critical role in determining the overall service workability of the joint. Filler metal can significantly change the weld metal properties with the proper balance of chemical constituents, which can alternatively act as constituent elements which were evaporated during high temperatures in welding. Filler wires used in the welding of corrosive resistance alloys such as Ni-Cr-Mo alloys are generally made with a higher alloy content in order to counteract the evaporation of alloying elements by high-temperature oxidation and to compensate for the alloying element-depleted region created due to segregation in the weld [2]. Guo et al. [3] investigated the welding of S960 HSLA steel using the same filler material composition. Although the impact toughness of the fused zone was higher, it did not gain sufficient strength and the failure occurred in the heat affected zone (HAZ).

Joining dissimilar materials is one of the crucial needs in various industries such as automobile, aerospace, oil and gas industry, and so on. The use of advanced materials is increasing in various industries with a significant challenge to withstand the welded structure in various components. Various situations in industry arise where the advantages of two different metals benefit from the outstanding performance, especially for economic benefits. Dissimilar material joints not only fulfill the criteria for service condition but also seem economical to minimize the usage of expensive metal in these places. These economic benefits necessitate the dissimilar welding of different metals. However, due to the different physical, mechanical,

 DOI: 10.1201/9781003435884-5

and metallurgical properties of the dissimilar metal to be joined, dissimilar metal welding is more crucial as compared to similar welding. The preference of filler metal is one of the challenging aspects which can either deteriorate or strengthen the weld joint properties. The constituent element of filler metal plays a crucial role on the strength of welded structure and corrosion resistance by phase transformation, and grain size effect [4]. In dissimilar metal joining, the formation of various intermetallic and solid solubility in various alloys are affected by the filler metal and welding procedure. The formation of brittle intermetallic compounds in dissimilar metal joints is vulnerable to cracking and corrosion attacks. Due to the variation in thermophysical properties of dissimilar material, it generates thermal stresses which leads to cracking in the material. Moreover, dissimilar metal weld joints are prone to galvanic corrosion in aggressive environments as they form a galvanic couple. However, the selection of filler metal can reduce the formation of cracks by selecting the filler wire that almost matches with the thermal properties of two different materials [5]. In the case of aluminium and steel dissimilar joining, controlling the intermetallic layer is the key index for mechanical properties and corrosion behavior of dissimilar assembly. The preference of filler wire for this dissimilar combination must possess a low melting point, good wettability with the base materials, and higher strength [5]. Inappropriate selection of filler metal electrodes for austenitic stainless steel pipes can cause hot cracking. Appropriate selection of filler wire controls the weld metal constituent elements and can help in preventing solidification cracking and reduces the chance of oxidation [6]. Improper selection of filler wire advances to various welding defects such as segregation of elements, dilution, secondary phase and intermetallic compound formation, and cracks. Moreover, the corrosion attack will be more intensive on the weld metal with the selection of lesser corrosive resistance filler wire than base metal [2]. Elemental migration or segregation is one of the major concerns in welding which influences the strength of weld metal. Carbon migration is observed in various weldments with the content of Cr-Mo in the weld [2]. The driving factor for carbon migration occurs by the elemental contrast in the weld region and base metal and more precisely by chromium content. Carbon has the tendency to segregate higher chromium content and forms chromium carbide mostly at the interface of the weld. Due to this, a carbon-depleted zone is formed and reduces the mechanical properties. This migration can be avoided with the preferable selection of filler wire during welding. To avoid carbon migration, nickel-based filler wires are utilized generally as the activity and diffusivity of carbon in nickel-based material are low [2].

2. SELECTION OF FILLER METAL

Filler material choice is critical for the service life of the welded structure, especially during the welding of dissimilar metals. The selection criteria must be practised using a combination of metallurgical principles and industrial service experience. The primary criterion of selecting the filler metal must be compatible with the material to be the welded and requires four major criteria: metallurgical compatibility, mechanical properties, physical properties, and corrosion resistance.

2.1. Metallurgical compatibility

The added filler material must be compatible with the dilution of both the workpiece material to be welded without cracking. The selected filler material must produce a sound weld with an admissible level of inclusion and porosity. The formation of various deleterious phases as well as the precipitation of new phases certainly alters the properties of solidified weld. Therefore, the selected filler metal must impart a stable weld metal microstructure after dilution of base material, when exposed to service conditions. For instance, selected filler metal may develop beneficial changes with exposure of service conditions, like work-hardening or age-hardening [7].

2.2. Physical properties

The physical properties of the weld zone as formed with the filler wire and dilution of base material must be compatible with that of the base material. For instance, the filler metal selected must have a melting range sufficiently lower to hinder solidification liquation cracking and to minimize partially melted regions and mixed zones in base material [7]. An instance of joining a high-melting material to a low-melting material is the joining of tantalum to a ferritic stainless steel (FSS) for a chemical process application. A dissimilar combination of tantalum and ferritic stainless steel was required for the assembly of the hot acid handling system. Tantalum shows superior corrosion resistance in the hot acid environment, which is the result of strong stable tantalum oxides. The extreme difference in the thermal expansion coefficient between tantalum (~7 x 10^{-6}/ºC) and ferritic stainless steel (~13 x 10^{-6}/ºC) can create a significant stress concentration at the interface of these joints during cooling and can create a crack. An active filler material, typically silver or copper, which forms a more stable oxide than the base material oxide was chosen for this dissimilar combination. The excellent joint was produced without any cracking with 72%Ag-28%Cu filler material [8].

The difference in thermophysical properties such as the coefficient of thermal expansion of the metal to be welded can change the heat flow in the welding and increase the residual stress on either side of the weld joint, transforming the HAZ to be weaker on the side of material with low coefficient of thermal expansion. The change in the chemistry of the material to be joined can lead to the formation of the deleterious brittle compound. In boiler heat exchanger tubes, weld between dissimilar combinations of austenitic stainless steel to ferritic stainless steel is often required. However, these steels are not much differed in chemistry and metallurgy but the difference in the thermophysical properties results in cracking during elevated temperature service. Various industrial application requires dissimilar combinations such as Ti-SS in the nuclear industry, and Al-steel in the automotive industry [9].

In order to reduce the carbon emission of automotive vehicles, the weight of the car body is minimized using the dissimilar combination of zinc-coated steel with lightweight aluminium alloy. This combination in the car body lowers the throughout weight of the vehicle and hence the reduction in fuel consumption which saves energy. However, the difference in thermophysical properties of aluminium and steel such as thermal expansion, and melting temperature, exhibits poor metallurgical compatibility [7]. One of the studies of C. Dharmendra et al. [10] on the dissimilar welding of coated steel to aluminium alloy with laser welding shows the usage of low melting point zinc-based filler material (Zn-15Al), which exhibits an excellent weld joint and the intermetallic layer was ~ 8 μm thick. Regardless of the formation of the intermetallic layer, the interface between the weld and steel was not a weakest point of dissimilar combination and the tensile failure occurred on the aluminium side. It is to be understood that the thickness of an intermetallic layer formed below 10 μm will not embrittle the weld joint. The coefficient of thermal expansion (CTE) of filler wire for dissimilar weld joints should be intermediate between the two different base materials to be joined so that it provides a minimum internal stress as a consequence of the thermal cycling of the dissimilar weld joint in the working environment [7].

2.3. Corrosion properties

The service life of the engineering component is mainly concerned with the corrosion resistance of the welded joint. The weld metal corrosion resistance must be equal to or higher than that of the base material to be joined. The corrosion resistance is mainly governed by the selection of filler material for the weldments. For the weld joint which is exposed to low-temperature aqueous environments and high-temperature corrosive environments, the overmatched filler material is preferred. This prevents preferential corrosion of the fusion zone. At lower working temperatures in an aqueous environment, the predominant corrosion problems that arise are galvanic pitting and crevice corrosion. With the proper selection of added filler electrode material, the potential detrimental attacks can be overcome. The filler metal chosen should be cathodic or close to the galvanic series of the base material to be joined [7]. During the joining of nickel-free stainless steel with high nitrogen content, the role of filler material critically affects the corrosion behavior as well as the mechanical response. The recent development of nitrogen alloyed stainless steels has found its application in the defence sector, cryogenic processes, pressure vessels and nuclear industries [7].

Nitrogen alloyed steels are prone to solidification cracking and produces porosity in the weld metal due to the improper selection of the filler material. Raffi et. al. [11] compared the different filler materials for the welding of nickel-free stainless steel with nitrogen content. Weld metal formed with 18Ni (MDN 250) filler possesses an unmixed region adjacent to the fusion boundary near weld metal. However, the weld metal formed with 13-8Mo (PH) filler metal exhibits higher hardness as well as enhanced pitting corrosion resistance in contrast to 18Ni (MDN 250) filler [11]. This improvement in pitting corrosion resistance is ascribed to the chemical composition of the 13-8Mo (PH) filler, where the existence of chromium and molybdenum generates a thin layer of stable passive film.

3. EFFECT OF FILLER MATERIAL ON THE GALVANIC CORROSION

During the welding of two materials with different compositions and metallographic structures, it is possible that an accelerated corrosion cell may form, as the two metals are in conjunction electrically with the presence of an electrolyte. This situation is ranked as a system called galvanic series, where the material ranked with their electrode potential in a given environment type of moist or aqueous such as saline solution, acid or soils [7]. This classification helps in determining the galvanic corrosion relationship during the selection of filler material for joining of same or different material with welding. Galvanic corrosion or sometimes referred as bimetallic corrosion and is associated with the current flow between the two materials associated in electrical contact in an electrolytic medium [7]. The galvanic series of various metals and alloys as per their potential difference is illustrated in Figure 5.1. Range of electrode potential of metals and alloys based on the actual measurement in flowing saline water is shown by an open bar in Figure 5.1. Some material exhibits active-passive behavior and is shown by two bars open and solid in Figure 5.1. The open bar indicates passive condition (i.e. no corrosion) of the electrode potential and the darkened solid blocks represents the range of the potential when the metal is in state of active condition such as localized corrosion like pitting and crevice.

During the dissimilar material welding, the role of filler material certainly affects the corrosion behavior of the two metals. The metals on the noble side of the galvanic series (Figure 5.1) do not corrode when welded to those above them toward the active side, however, the active material on the galvanic series will corrode [7]. The further away the two welded materials are in the list of galvanic series, the faster the corrosion will occur on the upper material toward the active side. Some metals such as austenitic stainless steel, and nickel-aluminium bronze possess both active-passive behavior. As the protective oxide layer film is intact with these metals, it will exhibit a high potential side in the galvanic series. However, if the formed oxide layer is broken or coupled to a metal more noble on the galvanic series such as titanium, then the material shifts to an active state as comparable to that of carbon steel on the galvanic series. Galvanic corrosion in weld metal is directly related to the current density. When the size of the anode material decreases, the specific current density affecting corrosion increases correspondingly. Preferably, the anodic material must be larger than the cathodic material. In case of dissimilar metal welding, the weld metal must be cathodic in a galvanic series with the base material [7].

The microstructure of the weld metal also has an influential role on galvanic corrosion. During the welding of similar metal also, there is the tendency to select filler metal slightly different from the base material composition due to the addition of deoxidizer, which subsequently creates segregation in the weld metal microstructure and tends to form an electrochemical difference that increases the chances of galvanic corrosion even for welding metals of the same composition. Some atmospheric conditions are often liable to pitting corrosion and the usefulness of the most economical and corrosion resistant material sometimes leads to the weld metal being the fragile zone in the joints because of micro-segregation. An illustration of the harsh atmospheric condition is the paper and pulp industry where an aqueous acid-chloride environment is a concern. Stainless steels with 3–6% molybdenum are commonly used in this kind of bleaching plants. However, preferential pitting corrosion in the weld is frequently observed when welding is made with a matching filler wire. This pitting corrosion attack on the weld zone is the subsequent effect of chromium and molybdenum micro-segregation within the dendritic structure during solidification. The formation of an inter-dendritic region constructs a dendrite core region of depleted chromium and molybdenum elements. As the chromium and molybdenum are the major alloying elements that impart resistance to pitting corrosion, the pitting will preferably attack the center of dendrite on the weld metal. However, the trick to avoid this situation occurs with the preference of over alloyed filler material. A filler material with adequate content of molybdenum such as ERNiCrMo-3 will offset the segregation affected pitting corrosion. A similar phenomenon also occurs in nickel alloy, where tungsten and molybdenum act as an effective element in providing pitting corrosion. The service condition where pitting crevice or galvanic attack are certain for nickel alloy is expected to adopt filler electrode material with a PRE (pitting resistance equivalent) higher than that of base metal. The relationship of PRE for the nickel alloy based on the composition of tungsten, molybdenum and chromium is as follows:

$$PRE = \%Cr + 1.7(\%Mo + \%W) \tag{5.1}$$

For several nickel alloys as listed in Table 5.1, it is recommended to use filler metal AWS ERNiCrMo-14 (UNS N06686) in order to provide better corrosion resistance [7].

Austenitic stainless steel (ASS) is one of the extensively practised steels in many industrial applications including offshore plants, LNG interior structures and building structure materials. It possesses superior mechanical and corrosion resistance properties; however, the weldment of ASS has inferior corrosion properties as compared to its base material. The corrosion properties of weldments depend on the alloying element as well as on the solidification behavior. The selected filler material plays a vital role in determining the properties of the weld metal. During the welding of austenitic stainless steel, the alloying element of the filler material especially Cr content determines the electrochemical behavior of the weld metal in an electrolytic solution [12][13].

In one of the studies by Kim et al. [14], where the joining of 316l ASS was carried out with three different filler metals of increasing order of Cr_{eq}/Ni_{eq} content as shown in Table 5.2. It was noticed that the weld metal with less than 3% delta ferrite was observed for the filler 1, which possesses ductility-dip cracking (DDC) which is one

TABLE 5.1
Pitting resistance equivalent number for Nickel-base alloys and recommended filler wire [7]

		Chemical composition, wt. %						
Alloy	**UNS No.**	**Fe**	**Ni**	**Cr**	**Mo**	**W**	**PRE**	**Filler wire**
C-276	N10276	6	57	15.5	16	3.9	49.33	AWS ERNiCrMo 14
C-22	N06022	4	56.5	21.4	13.4	3.1	49.45	
622	N06022	2	59.4	20.5	14.2	3.2	49.45	
686	N06686	1	57	20.5	16.3	3.9	54.84	
625	N06625	3	62	22	8.8	-	36.96	

TABLE 5.2
Chemical constituents of filler metal and the base metal (316L) [14]

Specimens	C	Si	Mn	P	S	Cr	Ni	Mo	Nb	Ti	Cr_{eq}	Ni_{eq}	Cr_{eq}/Ni_{eq}
Filler 1	0.02	0.6	1.1	0.028	0.005	16.2	13.8	2.6	0.02	0.03	19.7	14.9	1.3
Filler 2	0.02	0.6	1.1	0.029	0.004	18.2	11.2	2.5	0.02	0.03	21.6	12.3	1.7
Filler 3	0.02	0.5	1.2	0.028	0.003	21.1	9.3	2.5	0.02	0.03	24.3	10.8	2.2
316L ASS	0.02	0.6	1.5	0.031	0.004	18.3	11.5	2.4	-	-	-	-	-

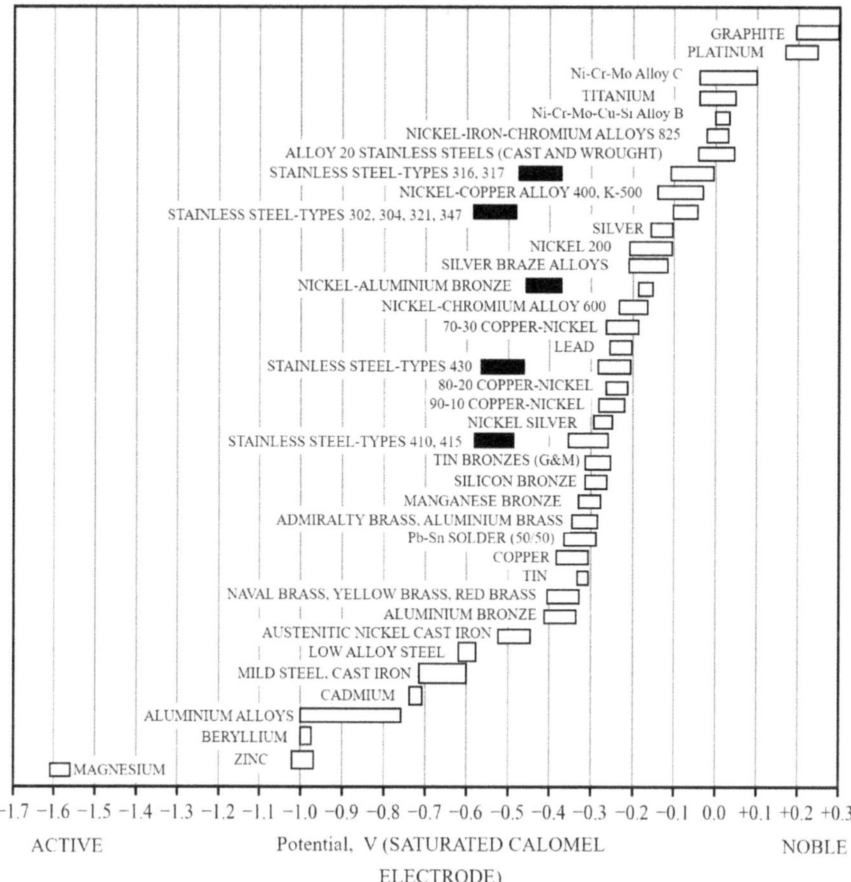

FIGURE 5.1 Galvanic series of metal and alloys exposed to saline water

Note: Metals and alloys are arranged as per the potential they showed in the flowing seawater. Certain alloys in low-velocity or poorly aerated water environments turns to active and demonstrate a potential near -0.5 volts as shown by the second darkened bar) [7].

of the forms of hot cracking. It is recommended to have at least 3 vol.% of delta ferrite in the weld zone to overcome hot cracking [15]. Ductility-dip cracking is usually realized in a reheated zone after multipass welding. In the austenite matrix, the presence of delta ferrite behaves as a pin to inhibit the grain boundaries from sliding when exposed to extreme restraint stress in the ductility-dip temperature range [16]. However, the delta ferrite less than 0.3 vol % is not effective in preventing ductility-dip cracking.

The weld region with the foremost Cr content (Filler 3) shows the topmost pitting potential and even the passivation range surpasses the base material. The filler metal with the lowest chromium concentration was degraded with localized and severe corrosion attacks on the austenite matrix.

4. EFFECT OF FILLER MATERIAL ON LIQUATION CRACKING

Weld solidification cracking is one of the subjects of deep concern in various metals and alloys. Weld solidification cracking is one of the issues of hot cracking phenomena which occurs after the solidification of the fusion zone. However, in the case of multipass welding, the cracking appears due to the reheating of weld metal and is called weld metal liquation cracking. Solidification cracking arises preferentially along solidification grain boundaries (SGBs) in the weld metal region [17]. Basically, two factors are responsible for weld solidification cracking (a) restrained strain due to mechanical and thermal conditions and (b) microstructure susceptible to cracking. The restrained factor occurs from the shrinkage phenomenon during solidification as the metal undergoes negative volume change during solidification [17]. An alternative governing factor that impacts restraint includes material properties, that is, base material as well as a selection of filler metal for welding. The permanent solution to restrict the solidification cracking is achieved by understanding the metallurgical basis of compositional modification of weld metal [17]. The chemical constituent analysis of weld metal is essential in controlling the vulnerability of cracking. The effect of discrete alloying elements or impurity in the filler metal is generally altered synergistically with the existence of other alloying elements. Prediction of cracking susceptibility of weld metal based on binary phase is uncertain. For this, it is essential to consider all the constituent elements when predicting the weld metal solidification cracking susceptibility. Various empirical relationships have been developed so far to anticipate the cracking susceptibility depending on chemical constituent. The equation established by bailey and jones for C-Mn steels is given as [17]:

$$CS_{TWI}\,(\text{wt.\%}) = 230C + 190S + 75P + 45Nb - 12.3Si - 5.4Mn - 1 \qquad (5.2)$$

It predicts cracking susceptibility as, for the value of CS_{TWI} <20, the weld metal is unaffected by cracking. This prediction also influences the anxious consequences of sulphur and carbon (above 0.08 wt%) as well as an advantageous result of silicon and manganese in steel. Generally, manganese tends to react with sulphur in the weld pool and forms a stable MnS compound. Thus, the presence of Mn reduces the effective quantity of sulphur responsible for cracking [18].

Based on the various published weldability data for steels, the weld metal solidification cracking susceptibility of steel based on Ni-equivalent and Cr-equivalent is shown in Figure 5.2. This diagram was constructed based on the weldability test data where restraints are high. Thus, higher value of Ni_{eq} can be accede to low restraint conditions [17]. With the increase in Ni-equivalent as well as Cr-equivalent, the transition region arises from a crack susceptible to crack-resistant composition. The transition region shows the change in the solidification behavior of weld metal from primary ferrite at a lower value of Ni_{eq} to primary austenite at a higher value [17].

Carbon manganese steel with a carbon % under 0.10 are immune to weld solidification cracking. For instance, carbon steel with a chemical constituent of 0.10C and 0.5 Mn has a nickel equivalent of 2.35, which lies in the crack-resistant region in Figure 5.2. In the case of austenitic stainless steel (300 series), weld solidification cracking can be insured by the ferrite content present in the weld metal.

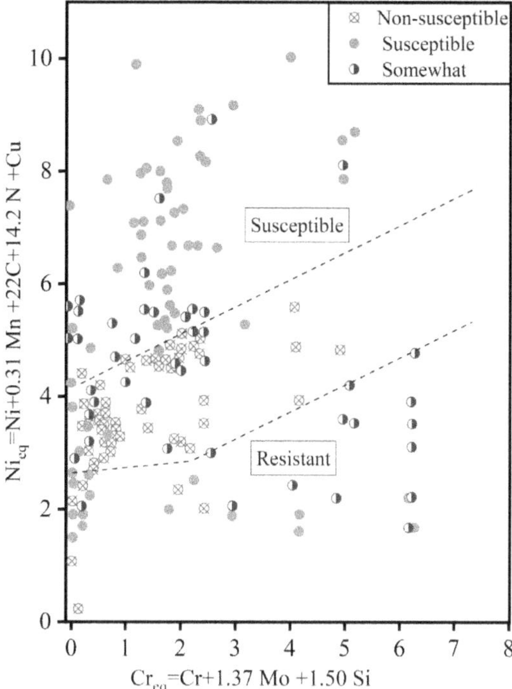

FIGURE 5.2 Weld metal solidification cracking of steel based on Ni_{eq} and Cr_{eq} [17]

Maintaining a ferrite number (FN) of more than 3 is normally required to avoid weld solidification cracking while with a high level of restraints [17]. Solidification mode is the primary measure to control the cracking. Solidification cracking susceptibility is quite low when the first phase to solidify is ferrite. The solidification mode and ferrite number for some of the metals is shown in Table 5.3. From the table it can be observed that stainless steels with a primary solidification phase as ferrite (bcc) are substantially more resistant to weld metal solidification cracking than other material which solidify as austenite (FCC) such as stainless steel, nickel-base alloys, and so on.

5. FILLER METAL EFFECT ON FUSION REGION

There are several metallurgical processes and phenomena that govern the microstructure and physical properties of weld joint. The key to achieve bearable welding joints in all the fusion welding process is the melting and solidification. Solidification pertains to the segregation and diffusion process that certainly occurs with local composition variation in the fused zone and influences the serviceability of weld joints. Weld metal fusion zones are the regions where complete melting and solidification occurs [17]. The microstructure obtained in the weld metal is the consequence of filler metal composition as well as solidification conditions [17]. A minute change

TABLE 5.3
Mode of solidification and ferrite number for nickel-base alloy and stainless steel [17]

Alloy	Solidification mode	Ferrite number
	Ni-base alloys	
Haynes 230W	A	NA
Hastelloy W	A	NA
Hastelloy X	A	NA
Hastelloy C-22	A	NA
Alloy 617	A	NA
Alloy 625	A	NA
	Stainless steels	
Austenitic, 304L	FA	6
Austenitic, 316L	FA	4
Austenitic, 310	A	0
Austenitic PH, A-286	A	0
Superaustenitic, AL6XN	A	0
Duplex, 2205	F	85
Duplex, 2507	F	75

in chemical composition often pertains to substantial changes in microstructure and properties of weld metal. The effectiveness of filler represents the fusion zone region. There are three types of fusion zone that have been described based on the usage of filler metal with that of substrate material [17]. These three types are Autogenous weld, Homogeneous weld, Heterogenous weld.

Autogenous welds are weld joints where the weld metal is formed by only melting and solidifying the two-base material without using filler material. This situation arises for the material whose thickness are minimal (<5 mm) and full penetration can be achieved without using filler material. Welding techniques with autogenous welds includes Laser Beam Welding (LBW), Electron Beam Welding (EBW), Gas Tungsten Arc Welding (GTAW), Plasma Arc Welding (PAW) and resistance welding. The fusion zone more or less possesses the identical composition as that of substrate material except for elemental loss during evaporation [17]. However, not all the material is feasible for joining without filler metal because of certain weldability issues.

Homogenous welding necessitates the need of additional filler material with the chemical constituents as that of the base material. These types of weld joints are required when the fused zone and substrate metal properties such as corrosion resistance and heat treatment response must be closely matched. Some of the examples of this type of weld include 316L SS base metal welding using 316L electrode wire as a means to maintain the same corrosion properties; similarly, ER10016-D2 electrode is adapted for joining of 4130 Cr-Mo steel, which requires a post-weld heat treatment in order to maintain uniform strength [17].

Heterogenous welding involves the use of filler metal whose chemical composition is distinct from the substrate material. On several occasions, suitable filler metal that matches the base material may not be available, or the desired joint properties may not be achieved with a similar filler metal composition. Several substrate materials have a composition that prevails poor weldability; hence, a dissimilar filler metal is accepted to achieve the desired properties. Few factors taken into consideration for the adaptation of different filler materials incorporates weld solidification cracking resistance, strength, corrosion resistance, weld defects (porosity) and operating characteristics of the filler metal. An example of a heterogeneous weld is the incorporation of 308L filler metal for welding 304L base material in order to achieve good weldability and corrosion resistance.

In heterogeneous welding situations, a transition zone is usually formed in the middle of a fully mixed fusion zone and unaffected substrate metal. However, the transition zone will be very small and undetectable if the change in compositional difference between the filler material and substrate is not significant, especially when the microstructure of the weld region and transition zone is similar. For the significant change in the compositional difference between the filler metal and base metal, as can be seen in the case of austenitic stainless steel or nickel alloy cladding to the plain carbon steel, the formation of the transition zone is evident and exhibits differential structure properties [17]. During the cladding of carbon steel with austenitic stainless steel filler wire 308L for corrosion protection, the transition zone reaches a composition where the austenite transforms to martensite on solidification. The formation of this narrow zone (~ 40 μm) of martensite phase adjacent to the fusion boundary possesses much higher hardness than the base metal or weld metal [17]. For some service conditions, a post-weld heat treatment region is required to temper this hardened transition zone.

5.1. Case Study 1

Di et al. [19] studied the effect of duplex stainless steel (DSS) filler wire to clad the X70 pipeline steel. Cladding of DSS over steels such as low carbon or low alloy steel is usually preferred in many industries, such as oil and gas and nuclear, due to the good strength, toughness, corrosion resistance, and requisite weldability. However, the joining with dissimilar material is complicated because of various dissimilar properties such as physical, chemical, and metallurgical. The formation of the transition zone in dissimilar weld metal possesses heterogeneous regions and exhibits an abrupt change in properties. This narrow transition zone with the martensite phase exhibits a hardened region and might be the reason for the premature failure of dissimilar material joining. Figure 5.3 shows the formation of a transition zone during the cladding of DSS 2209 to the X70 steel. The transition region microstructure consists of phases in the order of alpha ferrite + martensite, martensite, martensite + delta ferrite, martensite + austenite + delta ferrite with the change in the ratio of Cr_{eq}/Ni_{eq} from 0.36 to 2.55. The hardness of the transition region exhibits a sharp increase of microhardness value to 407 $HV_{0.2}$ at the center of the transition region, ascribed to the presence of martensite only, while the base material and clad layer possess hardness of 276 $HV_{0.2}$ and 259 $HV_{0.2}$ respectively.

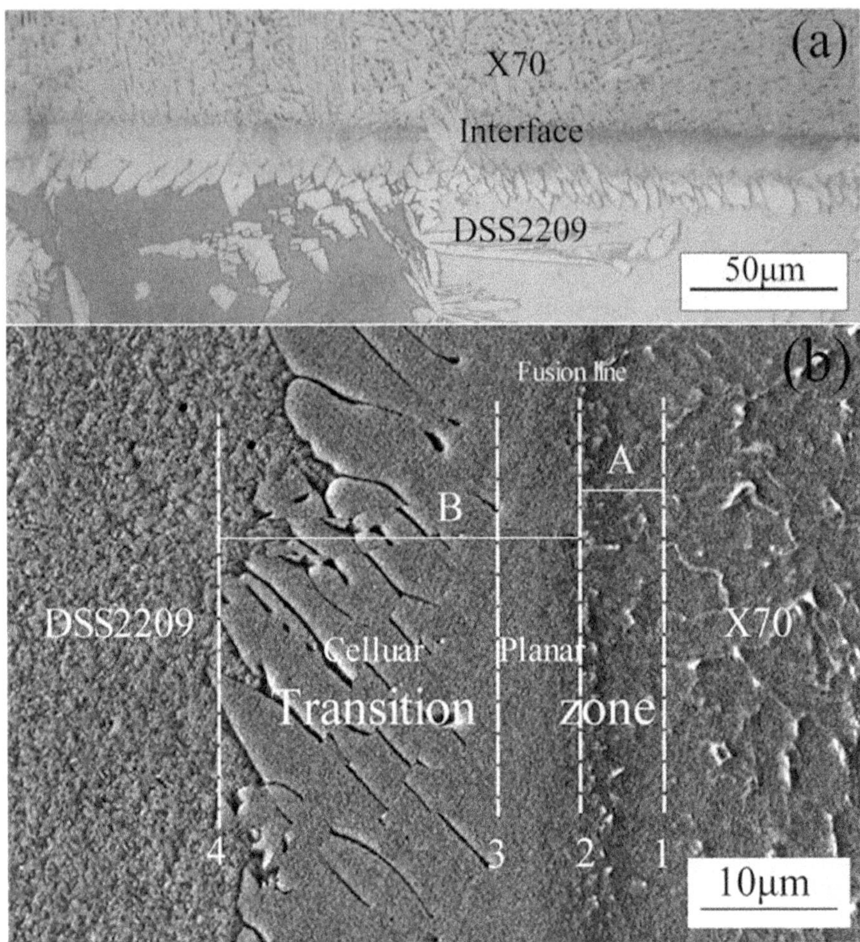

FIGURE 5.3 Formation of transition zone during the cladding of X70 pipeline steel with DSS as observed in (a) optical microscope (b) Scanning electron microscope

5.2. Case Study 2

The nuclear plant possesses a high risk of radioactive contamination and more concern for the safety during the natural disaster [20]. The pipelines used in nuclear industries are made from austenitic stainless steel as a result of their superior workability, mechanical performance, and corrosion resistance [21,22]. However, most of the pipelines in nuclear plants require a dissimilar joint of the austenitic stainless steel along with the low alloy steel. Inconel alloys such as 82 and 182 are mostly used as a filler material for producing the joint between ASS and low alloy steel, which is utilized in the nuclear plant at reactor coolant system. There have been accidents described due to the failure of dissimilar joints caused by water stress corrosion cracking [23]. The utilization of Inconel 52 and austenitic stainless steel has shown

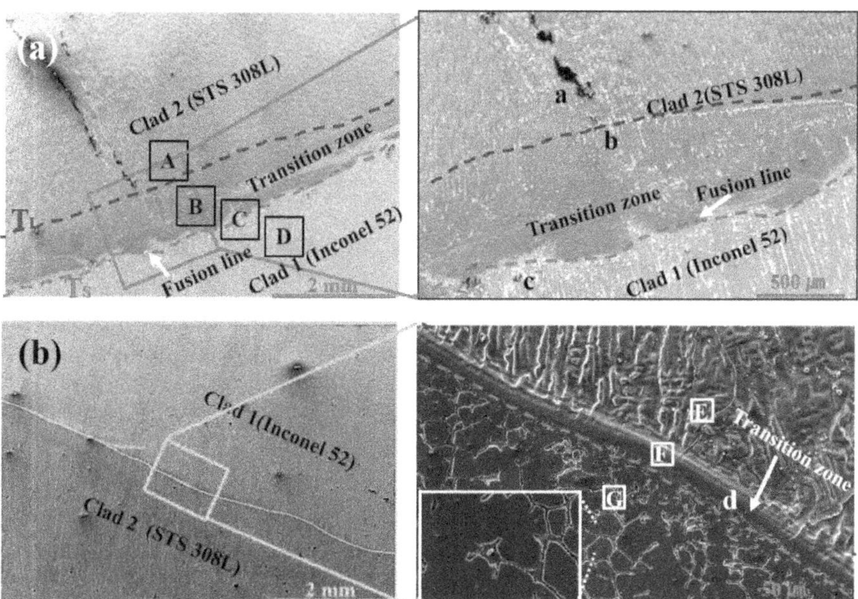

FIGURE 5.4 Formation of crack during the dissimilar clad of 308L ASS and Inconel 52 [26]

effective outcomes in achieving better stress corrosion cracking resistance. Moreover, hot cracking is a common problem in the joint of nickel-based alloy and austenitic stainless steel. It mostly occurred with the emergence of a brittle laves phase due to the high Ni/C ratio and the sensitivity of hot cracking increases with the increase in P and S content [24][25].

During the cladding of carbon steel with two distinct filler metals of Inconel 52 and ASS 308L [26], hot cracking is formed in the interfacial boundary between the fusion region of 308L and Inconel 52 as shown in Figure 5.4. The cracks were propagated into the austenitic stainless steel clad layer and few cracks were also seen underneath the transition zone. Some of the factors corresponding to crack formation were: (a) wide transition region in the interface amidst 308L and Inconel 52 dissimilar clad and (b) the formation of a thin film of eutectic-type laves phase along the grain boundary, which contains a high level of sulphur and phosphorous. Segregation of the phosphorous and sulphur is mostly present around the crack that appeared in the transition region. As a consequence, the development of a transition zone is critical which is mostly formed at the interface of dissimilar weld zones and it increases the susceptibility toward hot cracking.

6. DILUTION AND ITS EFFECT

The dilution effect is to be considered eminently when using a filler material with chemical constituents distinct from the substrate material to be joined. It can be described as the change in the chemical constituents of the weld zone due to the incorporation of filler metal into the substrate material during the melting process

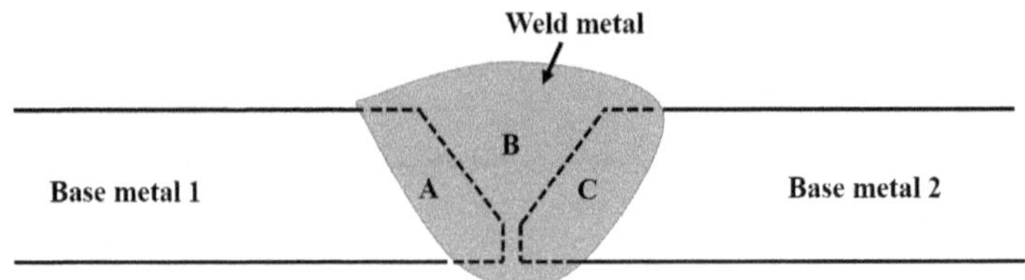

FIGURE 5.5 Schematic representation of the evaluation of dilution with filler metal

[17]. Dilution is sometimes considered inconvenient and it has to be controlled during the welding process. Modification of weld metal composition with the dilution of substrate material can certainly diminish the weld joint characteristics more than that obtained with the adaptation of a filler metal without dilution.

The specific situation where dilution is specifically unpleasant is the surface treatment process with the utilization of filler metals. In this surface treatment process, the filler metal is substantially different from the base metal composition and specifically selected to provide certain surface resistant characteristics such as abrasion, corrosion, or impact resistance. For the surface properties, stainless steel filler metals are practised as cladding material to protect the carbon steel from corrosion. However, the substantial dilution (~ 40%) can certainly decrease the chromium level to an extent where surface protection with the cladded layer is not useful for surface protection against corrosion [17].

Dilution can be described with filler metal and base metal as represented in Figure 5.5. It is expressed as the volume of the base metal melted to the volume of the fuse metal (including the base metal area represented in Figure 5.5 with the shaded region) as expressed in the following equation 5.3. For instance, a weld metal with 20% dilution contains 80% filler metal and 20% base metal.

$$Dilution\ \% = \frac{A+C}{A+B+C} \tag{5.3}$$

7. CONTROL OF SOLIDIFICATION CRACKING WITH FILLER METAL COMPOSITION

Solidification cracking is one of the major issues in the welding for several metals and alloys. However, solidification cracking can be avoided with the proper selection of filler material. The preference for filler material can supplement the chemical composition of the weld region and will avoid the susceptibility of the weld joints for cracking.

7.1. Aluminium Alloy

Aluminium alloys are prone to cracking; however, if the copper addition in the aluminium alloy is extended above 6%, cracking during solidification is predominantly hindered [27]. The weld region's chemical composition is regulated by the constituent elements of base metal, dilution ratio and filler metal selection. Consider an example of welding Al-3% Cu alloy, which is susceptible to weld metal cracking with 3% copper content. Choosing filler metal 2319 certainly reduces the solidification cracking, which has a copper concentration of 6.3% [27]. If the dilution is low, then weld metal certainly maintains copper more than 6% and reduces the chances of solidification cracking. Table 5.4. shows the choices of filler metal for aluminium alloy in order to avoid solidification cracking [27]. Some minor alloying element in aluminium such as Fe, and Si also affects the weld metal solidification cracking. For instance, the Fe-Si ratio significantly affects the weld solidification susceptibility cracking of 3004 and Al–Mg alloys. It was reported by Savage et al. that the decrease in Fe:Si ratio from 4.56 to 0.02 increases the hot cracking susceptibility. At a higher Fe:Si ratio (>2.0), the hot cracking susceptibility results from the formation of low melting β (Al–Mg) eutectic dispersed in grain boundaries. As the Fe content reduces, the formation of a semi-continuous network of (Mn, Fe) Al_6 at the grain boundary occurs and makes it brittle and more susceptible to hot cracking [27]. As the Fe content reduces to 0, the formation of (Mn, Fe) Al_6 changes to $MnAl_6$ and becomes less generous and more Mg_2Si forms in a continuous network at the grain boundary and increases the solidification cracking.

7.2. Carbon and Low Alloy Steel

Manganese content in the fusion zone has a remarkable impact on the solidification cracking. Manganese is often introduced for the formation of MnS compound. As MnS is formed, the formation of FeS can be hindered which is a key factor for

TABLE 5.4
Selection of filler metal to overcome cracking during solidification in aluminium alloy welding [27]

Base Metals → ↓	7000 (Al–Zn–Mg–Cu)	7000 (Al–Zn–Mg)	6000 (Al–Mg–Si)	5000 (Al–Mg)	2000 (Al–Cu)
2000 (Al–Cu)	NR	NR	NR	NR	4043 4145 2319
5000 (Al–Mg)	5356	5356 5556 5183	5356 5556 5183	5356 5556 5183	
6000 (Al–Mg–Si)	5356	5356 5556 5183	4083 4643 5356		
7000 (Al–Zn–Mg)	5356	5356 5556			
7000 (Al–Zn–Mg–Cu)	5356 5556				

NR- not recommended

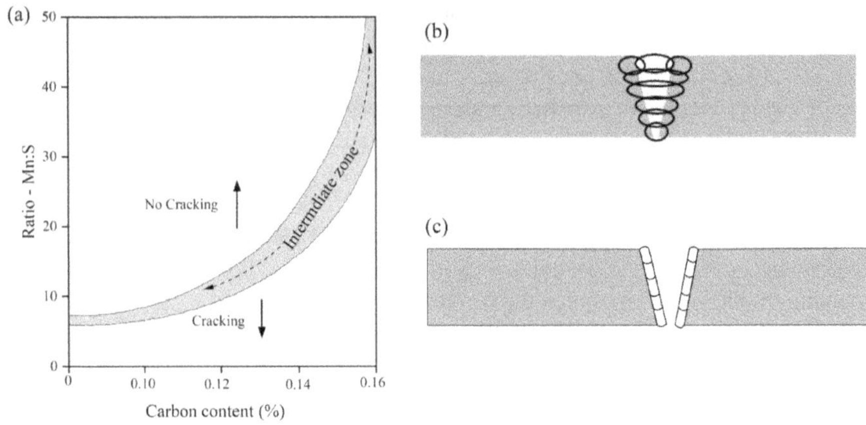

FIGURE 5.6 (a) Influence of carbon concentration and ratio of Mn:S on the cracking during solidification; (b) Multipass welding with highest dilution at the root pass; (c) Buttering of groove faces prior to welding of very high carbon steel

solidification cracking due to its low melting point. Figure 5.6(a) shows the role of the Mn-S ratio on the cracking of weld metal along with the carbon content. It can be observed that at low carbon levels, the solidification cracking occurred mainly due to the low ratio of Mn-S where more amount of FeS is formed which leads to cracking. However, at higher carbon content controlling alone Mn-S ratio is not beneficial for cracking; here formation of a brittle martensite phase can lead to post-solidification cracking of weld metal. Therefore, decreasing the carbon concentration in the fusion region is a more effective way to avoid solidification cracking. It can be attained by utilizing filler metal with a low level of carbon concentration. Moreover, a low carbon content filler metal is always recommended to weld a high carbon steel. This is due to the consideration that the first bead in multipass welding has higher dilution ratio (Figure 5.6 (b) and hence a higher carbon content than subsequent weld. Another way to decrease the carbon content in the fusion zone of high-carbon steel is to butter the groove faces with austenitic stainless steel filler rod prior to welding as illustrated in Figure 5.6 (c). For welding of cast iron, many times buttering with nickel filler material is carried out prior to welding. With this buttering, the surface layer remains in a ductile austenitic phase and the following welding can be carried out either with a stainless steel electrode or low carbon electrode. While selecting the filler metal for carbon steel to hinder cracking during solidification, the ratio of Mn-S should be carefully considered along with the carbon content.

8. PREDICTION OF WELD METAL CONSTITUTION

The prediction of weld region microstructure certainly varies with the usage of different filler materials. However, to understand the role of filler material, it is essential to acknowledge the fusion region microstructure. For instance, the weld region of austenitic stainless steel weld usually contains an austenitic (FCC) phase along with a small amount of δ-ferrite (BCC). δ-ferrite concentrations more than 10 vol % are prone to reduce the weld metal properties. However, δ-ferrite with less than 5 vol% may result in cracking during solidification [16]. The role of filler metal selection can certainly change the amount of austenite and ferrite in weld metal. The amount of the austenite and ferrite content in the weld metal can be predicted based on the constitution diagram.

Considerable effort has been made in the past years to predict the constitution of stainless steel fused regions. The most basic consideration in predicting the fusion zone constitution of stainless steel has been associated with the austenitic-ferritic alloy system. Strauss and Maurer in the year 1920 [16] established a nickel-chromium diagram with the correlation of nickel and chromium percentages. This chart allows predicting the respective phases present in the wrought steels and represents the stability lines among the phases such as austenite, martensite, and pearlite. However, the progress of research in this field leads to the development of a constitution diagram as a means to predict the weld metal microstructure.

Anton Schaeffler recognized that the prior research could be utilized and put into practice for welding. Schaeffler first proposed the constitution diagram in order to have a quantitative dependence between the chemical constituents and phase fraction (austenite or ferrite) of the fusion zone. He proposed a constitution diagram based on the combined effect from the Strauss-Maurer diagram that was applied for predicting wrought chromium-nickel steel and the Newell-Fleischman equation that could be practised for the welding. His constitution diagram focused on predicting the fusion zone microstructure depending on constituent elements. The Schaeffler's schematic representation contains chromium and nickel equivalent in the x and y axis respectively, with the specified weld metal microstructure phases. Elements which promote the ferrite phase are included in the chromium equivalent equation while the nickel equivalent equation contains all the elements which promote austenite phase formation. Schaeffler in the year 1949 established the concluding version of the constitution diagram after modification from its initial version and is as shown in Figure 5.7 (a) which is still in consideration. It was interesting to note that Schaeffler didn't consider putting the nitrogen in the nickel equivalent equation, even though it acts as a strong austenite stabilizer. This might be due to the complication of evaluating the exact amount of nitrogen composition in steel at that time. Due to the very low content of nitrogen in steels, it was not considered in the equation and a constant value of 0.06wt% was incorporated. This constitution diagram was designed based on the SMAW process. Further researchers have developed a diagram analogous to the Schaeffler diagram. Kakhovskii et al. [28] developed a schematic representation by modifying the Schaeffler diagram as shown in Figure 5.7 (b). This representative diagram considers the effect of alloying elements such as nitrogen, titanium, and vanadium.

DeLong et al. [29] introduced a constitution diagram in 1956 which focused particularly on 300-series austenitic stainless steel. The prediction of ferrite content was more precise for these particular austenitic stainless steels. It also considers the alloying effect of nitrogen on the fusion zone as it highly influences the microstructure. The DeLong diagram is represented in Figure 5.7 (c) and it differs from the

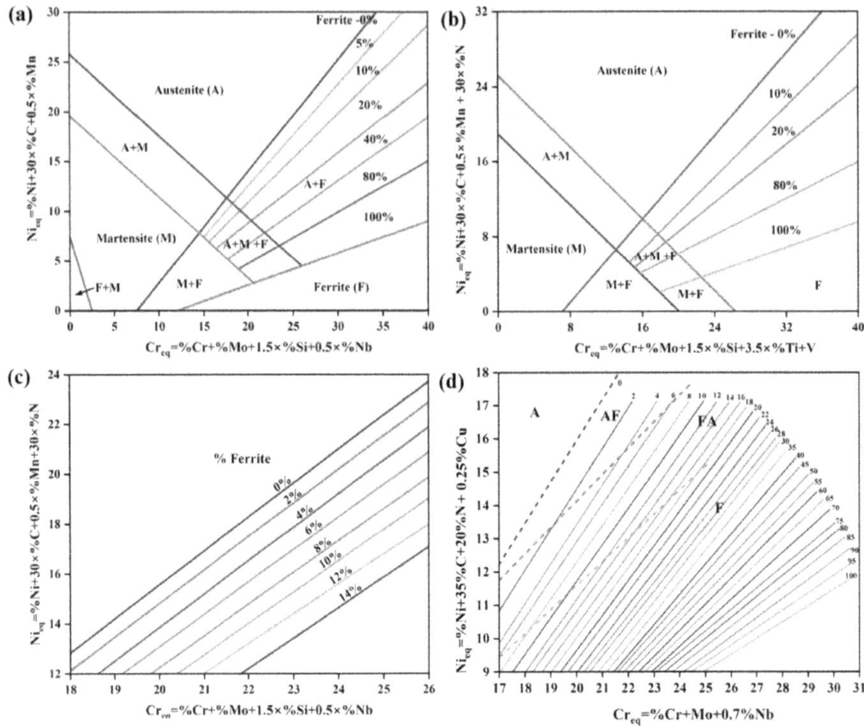

FIGURE 5.7 (a) Schaeffler diagram of 1949; (b) Kakhovskii diagram; (c) DeLong diagram of 1973 (d) WRC 1992 diagram [16]

Schaeffler diagram for the same area. The equation for nickel equivalent differs from Schaeffler's diagram and is given as:

$$Ni_{eq} = Ni + 0.5Mn + 30C + 30N \tag{5.4}$$

In DeLong's diagram, the inclination of isoferrite lines was greater than that of Schaeffler's diagram due to the discrepancies obtained among the evaluated and calculated ferrite fraction content of stainless steels. Further, there were modifications done by Long and DeLong in the constitution diagram. The major changes were the implementation of ferrite number (FN) in this illustration. FN measurements are in accordance with the magnetic behavior quantification, where δ-ferrite is basically ferromagnetic while austenite is non-magnetic.

In the decade of 1980, the subcommittee of the Welding research council commenced to emend and extend Schaeffler and DeLong's diagram in order to increase the accuracy for prediction of ferrite content in fused zone. Siewert et al. [30] put forward a predictive schematic representation known as *WRC-1988 diagram*. This diagram covered a wide span of composition starting with 0 to 100 ferrite number (FN) in contrast to that of the DeLong diagram that contains a narrow range of 0 to 18 FN. It also includes a boundary based on solidification modes such as A, AF, FA, and F. The chromium and nickel equivalent formula, as per the WRC-1988 diagram, is given as:

$$Cr_{eq} = Cr + Mo + 0.7Nb \tag{5.5}$$

$$Ni_{eq} = Ni + 35C + 20N \tag{5.6}$$

Studies on the influence of copper content on the ferrite phase became a subject of concern with the increased demand for duplex stainless steel, which contains copper up to 2%. It was proposed by the committee that the accuracy of FN may increase by adding a coefficient for copper content in the nickel equivalent equation of the WRC-1988 diagram. They proposed that a coefficient value of 0.25 to 0.30 be added to the copper coefficient. In 1992, a new diagram was proposed by Kotecki and Siewert, as shown in Figure 5.7(d), which is known as the WRC-1992 diagram [31] that incorporated a coefficient of 0.25 for the copper content in the nickel equivalent equation as:

$$Ni_{eq} = Ni + 35C + 20N + 0.25Cu \tag{5.7}$$

Although the WRC-1992 diagram was more accurate than the previous constitution diagram, it was noticed that the range of ferrite prediction was limited as compared to the Schaeffler diagram. Consequently, an extended version of this diagram was proposed in order to predict the ferrite number for dissimilar welding. However, the ferrite number prediction is accurate only when the composition of the fusion zone comes in the iso-FN line (0 to 100). At present, the WRC-1992 diagram is considered the utmost trustworthy for predicting the ferrite number in austenitic and duplex stainless steel. The limitation of this diagram lies in the absence of titanium in the chromium equivalent. Like niobium, titanium is also considered as a strong carbide former which can change the microstructure features by changing the phase composition by reducing the carbon concentration from the matrix. Titanium also acts as a strong ferrite stabilizer in the scarcity of carbon. Some of the austenitic stainless steel contains titanium and when it presents in an amount more than 0.2 wt.%, an additional factor of 2 to 3 in the chromium equivalent equation of the WRC-1992 diagram may enhance the precision for FN prediction.

8.1. Case Study—Selecting the Correct Filler Metal Based on Constitution Diagram

Joining of 320 stainless steel with 316L stainless steel requires special attention in order to choose the correct filler material to prevent cracking. The chemical constituents of the base metal, filler electrode and root pass are shown in Table 5.5. WRC-1992 diagram can be adapted to avoid fusion region solidification cracking

TABLE 5.5
Evaluation of weld metal composition formed by 316L and 320 base metal with E316L filler metal [16]

Elements	320 (A) wt.%	316L (B) wt.%	E316L (C) wt.%	70% of E316L (D = 0.70 *C)	15% of 316L (E = 0.15 *B)	15% of 320 (F = 0.15*A)	Weld metal composition (G = D + E + F)
C	0.02	0.02	0.03	0.021	0.003	0.003	0.027
Cr	20	17	18.5	12.95	2.55	3	18.5
Ni	34	12	12	8.4	1.8	5.1	15.3
Mo	2.5	2.3	2.3	1.61	0.345	0.375	2.33
Nb	0.30	0	0	0	0	0.045	0.045
Cu	3.5	0.2	0.2	0.14	0.03	0.525	0.695
N	0.02	0.02	0.06	0.042	0.003	0.003	0.048
Cr_{eq}	22.7	19.3	20.8				20.9
Ni_{eq}	35.97	13.1	14.3	-	-	-	17.4
FN	0	2.9	4.3				0

TABLE 5.6
Chemical composition of various filler metals along with Cr_{eq}, Ni_{eq}, and Ferrite Number [16]

	Nominal chemical composition (wt. %)									
Alloy	**C**	**Cr**	**Ni**	**Mo**	**Nb**	**Cu**	**N**	**Cr_{eq}**	**Ni_{eq}**	**FN**
E309L	0.03	23.5	13.5	0.2		0.2	0.06	23.8	15.7	10.5
E309L root	0.027	22.07	16.28	0.86	0.045	0.7	0.048	23.0	18.4	1.6
E309MoL	0.03	23.0	13.5	2.2		0.2	0.06	25.2	15.7	16.8
E309MoL root	0.027	21.65	16.28	2.26	0.045	0.7	0.048	23.9	18.4	3.5
E2209	0.03	22.5	9.0	3.0		0.1	0.15	25.5	13.1	35.4
E2209 root	0.027	21.3	13.2	2.82	0.045	0.6	0.111	24.2	16.5	9.0

for this dissimilar welding. Initially the welding engineer chose 316L filler metal; however, severe solidification cracking was observed. The usage of 316L filler metal certainly leads to no ferrite in the root pass as the 320 SS contains a high level of nickel and weld metal pickup nickel from 320 base metal. As discussed earlier, ferrite content with less than 5% in the fusion zone increases the chances of solidification cracking. With the use of a consumable electrode, dilution of 30% is usually achieved with 15% dilution from 320 base metal and other 15% from 316L base metal. Hence the chemical composition of root pass formed with 70% of 316L filler metal can be evaluated and is shown in Table 5.6. The chromium and nickel equivalent along with the ferrite number as per the WRC-1992 diagram is presented in Table 5.6. It can be understood that the use of filler metal 320 also will not be useful as the weld metal formed with it does not contain ferrite as shown in Table 5.5 and may have severe solidification cracking.

Several other filler metals can be considered such as 309L, 309LMo, 312, or 2209 duplex alloy. The chemical composition of these alloys along with the root pass composition with the dilution of base metal 316L and 320 SS base metal is shown in Table 5.5. 312 filler metal would be a good choice as far as solidification cracking is concerned. However, it contains 0.08 wt.% carbon which certainly reduces the corrosion resistance in the joint as both base metals contain very low carbon than this. To ensure the other filler material, the same sort of analysis has been done with the assumption of 30% dilution and has been incorporated in Table 5.5. Although E309L contains a sufficient amount of ferrite (FN -10.5), the E309L root pass contains less ferrite because of the dilution and can possess solidification cracking.

The root pass obtained using 309MoL has sufficient ferrite and looks acceptable; however, this is not the case. For this, refer to the WRC-1992 diagram (Figure 5.7 (d) where dotted lines represent the area under the solidification mode of A, AF, FA or F. In the WRC-1992 diagram, 309MoL root pass lies in the region of AF solidification mode. It can be well understood that the primary austenite solidification mode (AF) is sensitive to solidification cracking whereas the primary ferrite solidification mode (FA) is non-sensitive to solidification cracking. It is usual to have an FN of more than

3 or 4 to be free from solidification cracking. However, highly alloyed stainless steel requires a ferrite number of more than 5.

It can be concluded from the analysis, as shown in Table 5.6, that filler metals 309L and 309MoL are not safe and are sensitive to solidification cracking during the dissimilar welding of 320 and 316L stainless steel. However, 2209 filler metal is the best choice, as its root pass contains sufficient ferrite number (9) with the solidification mode of FA. A slightly higher amount of molybdenum content in the weld metal using 2209 filler wire makes it more appropriate. The corrosion resistance and tensile strength of 2209 filler wire also appeared to be higher than either of the base metal 320 and 316L stainless steel. It is recommended to use 2209 filler wire; however, consideration should be taken where the weldment is intended to service conditions where ferrite is detrimental, such as (a) urea production unit, (b) high-temperature service environment where the formation of sigma phase occurs, and (c) cryogenic environment where toughness is suppressed.

REFERENCES

[1] C. Schneider, W. Ernst, R. Schnitzer, H. Staufer, R. Vallant, N. Enzinger, Welding of S960MC with undermatching filler material, Weld. *World*. 62 (2018) 801–809. https://doi.org/10.1007/s40194-018-0570-1

[2] K. Devendranath Ramkumar, N. Arivazhagan, S. Narayanan, Effect of filler materials on the performance of gas tungsten arc welded AISI 304 and Monel 400, Mater. *Des.* 40 (2012) 70–79. https://doi.org/10.1016/j.matdes.2012.03.024

[3] W. Guo, D. Crowther, J.A. Francis, A. Thompson, Z. Liu, L. Li, Microstructure and mechanical properties of laser welded S960 high strength steel, *Mater. Des.* 85 (2015) 534–548. https://doi.org/10.1016/j.matdes.2015.07.037

[4] B. Yelamasetti, S. Kumar, B.S. Babu, T.V. Vardhan, V.R. Gunda, Effect of filler wires on weld strength of dissimilar pulse GTA Monel 400 and AISI 304 weldments, *Mater. Today Proc.* 19 (2019) 246–250. https://doi.org/10.1016/j.matpr.2019.06.759

[5] karim A. Md., P. Yeong-Do, A review on advances in friction welding of dissimilar metals, *J. Weld. Join.* 38 (2020) 8–23. https://doi.org/10.1007/978-981-19-0676-3_15

[6] A. Kumar, P.K. Dixit, Investigating the effects of filler material and heat treatment on hardness and impact strength of TIG weld, *Mater. Today Proc.* 26 (2019) 2776–2782. https://doi.org/10.1016/j.matpr.2020.02.578

[7] American welding society, Welding Handbook-Materials and applications, 1991.

[8] EWI, Joining Tantalum to a Ferritic Stainless Steel, (2014). https://ewi.org/joining-tantalum-to-a-ferritic-stainless-steel/

[9] B. Shanmugarajan, G. Padmanabham, Fusion welding studies using laser on Ti — SS dissimilar combination, *Opt. Lasers Eng.* 50 (2012) 1621–1627. https://doi.org/10.1016/j.optlaseng.2012.05.008

[10] C. Dharmendra, K.P. Rao, J. Wilden, S. Reich, Study on laser welding-brazing of zinc coated steel to aluminum alloy with a zinc based filler, *Mater. Sci. Eng. A.* 528 (2011) 1497–1503. https://doi.org/10.1016/j.msea.2010.10.050

[11] R. Mohammed, G.M. Reddy, K.S. Rao, Effect of filler wire composition on microstructure and pitting corrosion of nickel free high nitrogen stainless steel GTA welds, *Trans. Indian Inst. Met.* 69 (2016) 1919–1927. https://doi.org/10.1007/s12666-016-0851-6

[12] M. Keddam, H. Takenouti, D. Thierry, Modelling of the passivation mechanism of Fe-G binary alloys from ac impedance and Behaviour of Fe-Cr alloys in 0.5 M H2SO4

with an addition of chloride, *Electrochim. Acta.* 42 (1997) 1595–1611. https://doi.org/https://doi.org/10.1016/S0013-4686(96)00321-0

[13] I. Annergren, M. Keddam, H. Takenouti, D. Thierry, Modelling of the passivation mechanism of Fe-Cr binary alloys from ac impedance and frequency resolved rrde—I. Behaviour of Fe-Cr alloys in 0.5M H2SO4, *Electrochim. Acta.* 41 (1996) 1121–1135. https://doi.org/https://doi.org/10.1016/0013-4686(95)00463-7

[14] Y.H. Kim, D.G. Kim, J.H. Sung, I.S. Kim, D.E. Ko, N.H. Kang, H.U. Hong, J.H. Park, H.W. Lee, Influences of Cr/Ni equivalent ratios of filler wires on pitting corrosion and ductility-dip cracking of AISI 316L weld metals, *Met. Mater. Int.* 17 (2011) 151–155. https://doi.org/10.1007/s12540-011-0221-1

[15] V. Shankar, T.P.S. Gill, S.L. Mannan, S. Sundarlsan, Solidification cracking in austenitic stainless steel welds, *Sadhana - Acad. Proc. Eng. Sci.* 28 (2003) 359–382. https://doi.org/10.1007/BF02706438

[16] J.C. Lippold, D. Koteck, *Welding Metallurgy and weldability of stainless steel*, New Jersey, 2005.

[17] J.C. Lippold, *Welding Metallurgy and Weldability*, 2014. https://doi.org/10.1002/9781118960332

[18] H. Nakagawa, F. Matsuda, T. Senda, Effect of sulfur on solidification cracking in weld metal of steel (Report 1):: Fundamental Investigation on Sulphides in Fe-S Binary Alloy Steel, Trans. *Japan Weld. Soc.* 5 (1974) 39–44. https://doi.org/http://dl.ndl.go.jp/info:ndljp/pid/10944677

[19] X. Di, Z. Zhong, C. Deng, D. Wang, X. Guo, Microstructural evolution of transition zone of clad X70 with duplex stainless steel, *Mater. Des.* 95 (2016) 231–236. https://doi.org/10.1016/j.matdes.2016.01.087

[20] C. Lin, T. Su, K. Wu, Effects of parameter optimization on microstructure and properties of GTAW clad welding on AISI 304L stainless steel using Inconel 52M, *Int J Adv Manuf Technol.* 79 (2015) 2057–2066. https://doi.org/10.1007/s00170-015-6875-y

[21] R.M. Molak, K. Paradowski, T. Brynk, L. Ciupinski, Z. Pakiela, K.J. Kurzydlowski, International journal of pressure vessels and piping measurement of mechanical properties in a 316L stainless steel welded joint, *Int. J. Press. Vessel. Pip.* 86 (2009) 43–47. https://doi.org/10.1016/j.ijpvp.2008.11.002

[22] S.L. Jeng, H.T. Lee, W.P. Rehbach, T.Y. Kuo, T.E. Weirich, J.P. Mayer, Effects of Nb on the microstructure and corrosive property in the Alloy 690 – SUS 304L weldment, *Mater. Sci. Eng. A 397.* 397 (2005) 229–238. https://doi.org/10.1016/j.msea.2005.02.042

[23] R. Celin, F. Tehovnik, Degradation of A Ni-Cr-Fe alloy in a pressurised-water nuclear power plant, *Mater. Technol.* 45 (2011) 151–157. https://doi.org/UDK 621.311.25:620.193:669.14

[24] H. Naffakh, M. Shamanian, F. Ashrafizadeh, Weldability in dissimilar welds between Type 310 austenitic stainless steel and Alloy 657, *J. Mater. Sci.* 43 (2008) 5300–5304. https://doi.org/10.1007/s10853-008-2761-4

[25] H.A. Chu, M.C. Young, H.C. Chu, L.W. Tsay, C. Chen, H.A. Chu, C. Chen, M.C. Young, H.C. Chu, L.W. Tsay, The Effect of Nb and S segregation on the solidification cracking of Alloy 52M weld overlay on CF8 stainless, *J. Mater. Eng. Perform.* 23 (2014) 967–974. https://doi.org/10.1007/s11665-013-0812-8

[26] Y. Kim, H. Nam, J. Lee, C. Park, B. Moon, D.G. Nam, S.H. Lee, N. Kang, Hot-cracking resistivity of dissimilar clads using Inconel 52 and 308L stainless steel on carbon steel, *J. Nucl. Mater.* 533 (2020) 152103. https://doi.org/10.1016/j.jnucmat.2020.152103

[27] S. Kou, *Welding Metallurgy*, Second, New Jersey, 2003. https://doi.org/10.22486/iwj.v4i3.150243

[28] N.I. Kakhovskii, V.N. Lipodaev, G.V. Fadeeva, The arc welding of stable austenitic corrosion-resisting steels and alloys, *Avtom. Svarka.* 33 (1980) 55–57.
[29] W.T. DeLong, G.A. Ostrom, E.R. Szumachowski, Measurement and calculation of ferrite in stainless-steel weld metal, *Weld. J.* 35 (1956) 521s-528s.
[30] A. Siewert, C.N. Mccowan, D.L. Olson, Ferrite number prediction to 100 FN in stainless steel weld meta, *Weld. J.* 67 (1988) 289s-298s.
[31] D.J. Kotecki, T.A. Siewert, WRC-1992 constitution diagram for stainless steel weld metals: A modification of the WRC-1988 diagram, *Weld. J.* 71 (1992) 171s-178s.

6 Effect of Welding Parameters on Structure-Property Correlation

Santosh K. Gupta, Vipin Tandon, and Himanshu Vashishtha

1. INTRODUCTION

Welding is an inevitable phenomenon employed for joining materials in diverse sectors, including gas, chemical, and oil industries, among others [1]. The primary goal of this joining process is to ensure that the assembled components possess identical properties to the parent material. In other words, the weldment characteristics (mechanical and corrosion) should be on par with the base metal (BM) [2]. Depending on the specific application, welding can be performed using either similar or dissimilar BMs. In the contemporary industrial landscape, there is a growing demand for the use of innovative materials integrated across various domains, driven by considerations of cost-effectiveness, production expenses and overall maintenance. Consequently, dissimilar metal welding (DMW) emerges as a pivotal process that not only imparts the required characteristics but also enhances flexibility in various service conditions [3]. In addition to simply meeting the diverse needs of different service conditions, including aspects such as corrosion and heat resistance, as well as magnetic properties, the application of DMW can yield significant cost savings by reducing reliance on novel and costly materials. This, in turn, contributes to a lowering in the overall cost of the final product [4].

Fusion-based welding methods like laser beam welding (LBW), electron beam welding (EBW), and gas tungsten arc welding (GTAW) rely on the process of melting and then solidifying the weld metal (WM). These techniques may encounter certain challenges, such as heat-affected zone (HAZ) softening and a considerable HAZ width, which can potentially result in failures under dynamic operating conditions. Nevertheless, the careful selection of process parameters can substantially reduce these problems [5]. A solid-state joining technique like friction stir welding (FSW), lowers production expenses by eliminating defects, the need for shielding gas, and costly weld preparation. Furthermore, this method generates superior joints characterized by a more uniform microstructure and enhanced mechanical properties compared to traditional welding techniques [6].

Various categories of metallic materials, encompassing both ferrous materials like stainless steels (SSs) and non-ferrous materials like copper, aluminum, zinc,

DOI: 10.1201/9781003435884-6

and others, find extensive application in diverse joining processes. Within this array of materials, the SSs, in particular, stand out as notable engineering materials due to their exceptional attributes, including remarkable resistance to corrosion, robust high-temperature strength, resilience at cryogenic temperatures, and favorable characteristics for fabrication [7]. The utilization of SSs has experienced a significant upsurge in sectors such as petrochemicals, nuclear power generation, pressure vessel manufacturing, heat exchanger production, as well as the paper and pulp industries. This growth is primarily attributable to the extended lifespan and durability of components and structures constructed with stainless steels [8]. To confer stainless properties upon steels, it is necessary to have at least 11% (by weight) of chromium. At this critical chromium level, the steel surface can generate a tenacious, self-repairing chromium oxide layer, particularly under relatively neutral environmental conditions [9].

The welding processes and their associated parameters, including welding current, voltage, and welding speed, assume a pivotal role due to their profound impact on crucial properties such as the mechanical and corrosion characteristics of welded joints [10]. Moreover, the heat input (HI) of welding process is meticulously determined as per the parameters illustrated in Eq. 1 [11]:

$$HI = \eta VI/v \qquad \text{(Eq. 1)}$$

wherein "I" represents the welding current (measured in amperes, A), "v" signifies the welding speed (measured in millimeters per second, mm/s), "V" denotes the arc voltage (measured in volts, V), and "η" symbolizes the welding efficiency. The following elucidates how the geometry of the weld is affected by welding parameters:

(i) **Welding Current:** During the process of welding, the current responsible for creating the arc for melting and heating, holds a critical role. Notably, the HI experiences substantial variation with changes in current (as depicted in Eq. 1). An increase in current leads to a corresponding rise in HI, subsequently affecting the depth of penetration in the weldment, and width of weld bead in the weldment, the extent of the WZ, and the HAZ, all of which tend to increase concomitantly [12].

(ii) **Welding speed:** The welding speed is a pivotal factor directly dictating the quality of the weld bead's geometry. As per Equation 1, a higher welding speed leads to in a decrease in HI, potentially leading to inadequate penetration. Conversely, a slower welding speed elevates the HI, allowing grains more time to absorb the heat, which can result in a non-uniform columnar structure. Therefore, it is imperative to maintain an optimal welding speed to attain a more refined, uniform, axial, and homogeneous structure [13].

(iii) **Voltage:** Voltage is the electrical potential difference between the welding wire's tip and the surface of the molten weld pool, and it is crucial in determining the arc length. In comparison to low welding voltages, higher welding voltages tend to produce flatter, wider, and less deeply penetrated welds. Therefore, to achieve the desired depth of penetration, it is essential to maintain an optimal welding voltage [14].

Welding is a fundamental and widely used process in various industries, from automotive and aerospace to construction and manufacturing, and the SSs are the widely used materials. The quality and integrity of the SSs welded joints are crucial for the performance and safety of the final products. Achieving the desired mechanical and metallurgical properties in a welded structure depends on a complex interplay of factors, with welding parameters being a key determinant. Hence, this chapter seeks to delve into the intricate relationship between welding parameters and the resulting structure-property correlation, shedding light on the vital role that process control and optimization play in welding technology in similar and dissimilar welding of various types of SSs (such as austenitic, ferritic, duplex, and martensitic).

2. INFLUENCE OF WELDING PARAMETERS ON SIMILAR AND DISSIMILAR WELDING OF SS

The SSs have three main types of microstructures: ferritic, austenitic, and martensitic. Accordingly, the SSs can be broadly categorized into four main classes, (i) austenitic, (ii) ferritic, (iii) duplex, and (iv) martensitic SSs, based on these three primary microstructures (Figure 6.1).

The chemical composition of both the BM and the filler electrodes is the governing factor determining the solidification mode and microstructure of the WZ. These characteristics can be classified into four distinct groups, as outlined in Table 6.1.

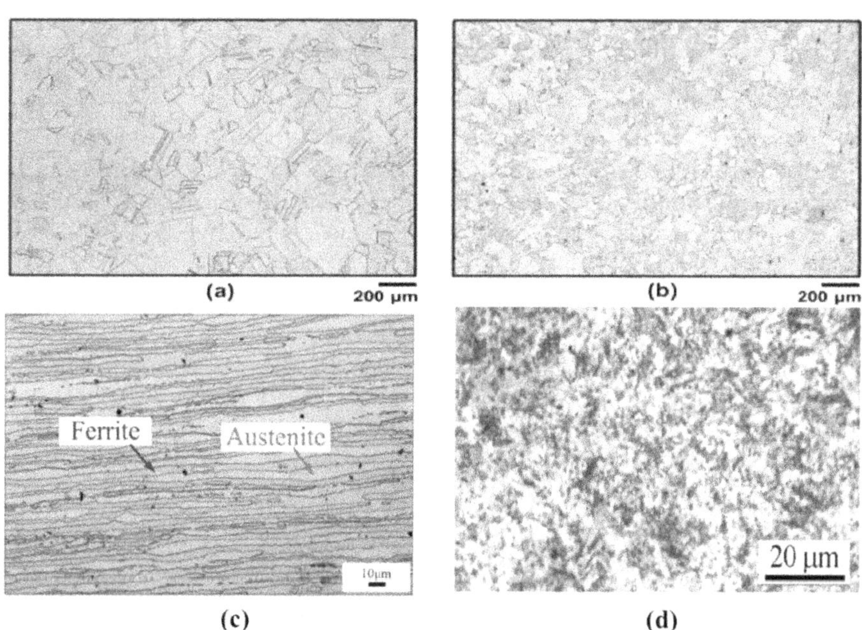

FIGURE 6.1 Optical microstructures of (a) 304L ASS [15] (b) 439 FSS [15] (c) 2205 DSS [16] (d) 420 MSS [17]

TABLE 6.1
Solidification types, reactions, and resultant microstructures [18]

Solidification type	Reaction	Microstructure
Austenite (A)	$L \rightarrow L + A \rightarrow A$	Fully austenite, well-defined solidification structure
Austenite-Ferrite (AF)	$L \rightarrow L + A \rightarrow L + A + (A + F)_{eut} \rightarrow A + F_{eut}$	Ferrite at cell and dendrite boundaries
Ferrite-Austenite (FA)	$L \rightarrow L + F \rightarrow L + F + (F + A)_{per/eut} \rightarrow F + A$	Skeletal and/or lathy ferrite resulting from ferrite to austenite transformation
Ferrite (F)	$L \rightarrow L + F \rightarrow F \rightarrow F + A$	Acicular ferrite or ferrite matrix with grain boundary austenite and Widmanstatten side plates

2.1. Similar welding of SSs

In this section, the influence of different welding parameters on the welding characteristics of similar welding of stainless steels is reported.

2.1.1. Austenitic stainless steel (ASS)

Several welding methods and their corresponding parameters were used to assess the welding characteristics in the context of welding similar austenitic stainless steels (ASSs). The study emphasized on microstructural and mechanical characteristics of joints created from 304L ASS using three distinct welding techniques: laser, tungsten inert gas (TIG), and laser-TIG hybrid is investigated in [19]. The utilization of these techniques resulted in varying levels of HI, which in turn exerted a substantial influence on the microstructural and mechanical features of the joints. Figure 6.2 (a–c) depict the distinct differences in the width and fusion area of the joints created through these various techniques, primarily attributed to differences in both HI and welding speed. Notably, the TIG joint (J_T – shown in Figure 6.2 (a)) exhibited a greater width and fusion area compared to the other techniques. This discrepancy can largely be attributed to the higher HI employed in TIG welding process. Figure 6.2 (d–i) presents the microstructures of all three joints. In the optical micrographs shown in Figure 6.2 (d–f), the region near the weld junction of these joints can be seen. There is a clear distinction between the BM and the WZ in the laser joint (JL) and hybrid joint (JH). Columnar dendrites are observable in these joints, extending from the fusion boundary (FB) to the weld centerline. Notably, no transition zone or HAZ is evident near the FBs in both the J_L and J_H joints. In contrast, the J_T joint exhibits clear transition zones and HAZ near the FBs. These differences in microstructural characteristics indicate variations in the welding process parameters and their consequent effects on the joint's structure. In Figure 6.2 (g–i), typical micrographs showcase the WZ of the three joints, all characterized by a common feature – a dark dendritic δ ferrite structure embedded within an austenite matrix. In each of these joints, the solidification mode is recognized as ferritic-austenitic (FA), and the quick cooling during

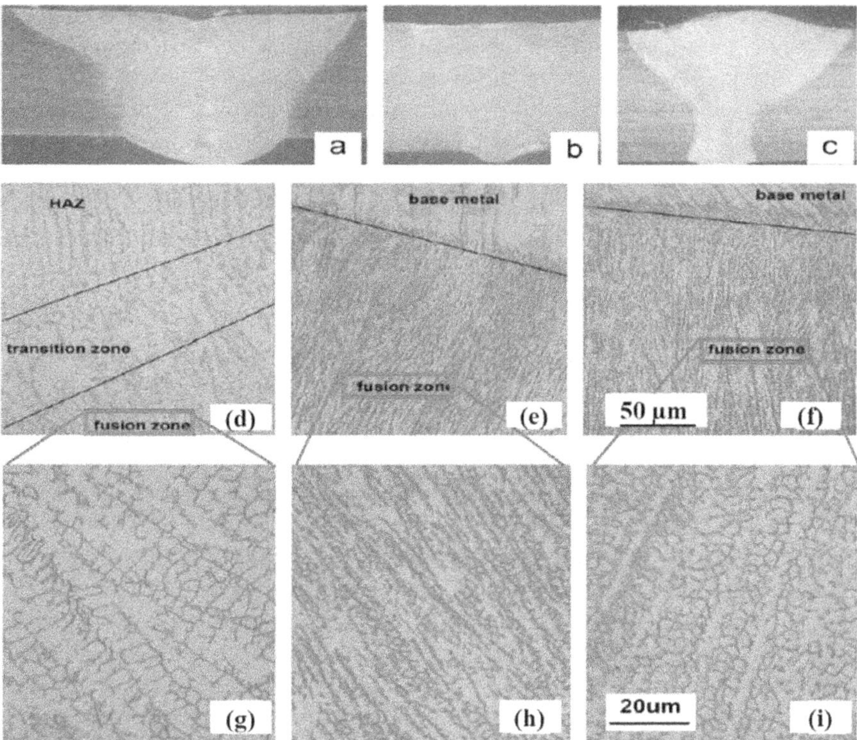

FIGURE 6.2 Macrostructure of 304L ASS of (a) J_T (b) J_L (c) J_H joint; Microstructure of (d, g) J_T joint (e, h) J_L joint (f, i) J_H joint. Note: J_T- TIG joint, J_L-Laser Joint, J_H- hybrid joint [19]

welding hinders adequate time for the phase transformation to finalize, leading to the presence of δ ferrite in the weld zone WZ. The J_T joint, due to its higher HI and slower cooling rate, experiences a reduction in tensile strength. In contrast, the J_L and J_H joints, with lower HI and faster cooling rates, exhibit reduced WZ and HAZ width, as the weldment doesn't have adequate time for these changes to occur.

When welding 304L ASS joints using the GTAW technique with varying HI [20], several noteworthy observations come to light. In all the weldments, the HAZ displays a noticeable coarsening of grains, with the grain size in the HAZ increasing in tandem with the rise in HI. This phenomenon is mirrored in the WZ, where both the inter-dendritic spacing and the average dendrite length increase as HI escalates. The rationale behind this trend is that a lower HI results in a relatively higher cooling rate, which impedes dendrite growth, while a higher HI affords dendrites more time to develop. This variation in grain size emerges as the primary contributor to the discernible alterations in the tensile properties of the weldments produced with varying HI conditions. Smaller dendrite size in WZ, associated with lower HI, translates into enhanced tensile strength and ductility compared to weldments with higher HI [20].

Furthermore, the GTAW of low-nickel and 304 ASS, employing varying levels of HI (low, medium, and high), produces shifts in microstructural and electrochemical properties [21]. In all these joints, the WZ is depicted the presence of vermicular and lathy δ ferrite. Lower HI leads to a closer inter-dendritic spacing and reduction in dendritic arm length in the weld microstructure. Conversely, increasing HI results to an expansion in HAZ width. And, weldments with lower HI exhibit enhanced mechanical properties, including increased microhardness and tensile strength, primarily attributable to a higher proportion of δ ferrite in the WZ under lower HI conditions. Electrochemical assessments reveal that weldments with lower HI exhibit reduced intergranular corrosion resistance but enhanced pitting and impedance corrosion resistance [21]. The microstructural and mechanical characteristics of 304L ASS welded by gas metal arc welding (GMAW) are significantly influenced by welding parameters, particularly welding current, and welding speed [7]. Two different joints, W1 and W2, with different production parameters, show different characteristics. When compared to W2, formed at a lower welding speed and a higher welding current, W1, generated with a lower welding current and a higher welding speed, leads to the FA solidification mode in both instances. Consequently, both joints exhibit a lathy and skeletal δ ferrite composition inside an austenite matrix in the WZ. It is noteworthy that the WZ of W1 has a higher δ ferrite content than its counterpart. Because of its finer grain structure, the WZ of W1 exhibits superior hardness and tensile strength when compared to other regions, such as the FB, HAZ, and BM [7].

Apart from the parameters associated with fusion welding techniques, the parameters governing solid-state welding techniques, such as FSW, also wield a significant influence over the microstructural and mechanical characteristics of ASS. Notably, the welding speed plays a pivotal role in shaping the microstructure within the WZ, consequently impacting both the strength and toughness of the joint. As welding speed increases, there is a discernible reduction in grain size within the weld stir zone, leading to a direct enhancement in mechanical properties (hardness and tensile strength). In terms of overall strength, toughness, and productivity, higher welding speeds offer clear advantages. However, it's important to note that exceeding a specific threshold in welding speed can result in inadequate material plasticization and coalescence, potentially leading to the development of internal tunnel defects [22]. During the FSW of 304 ASS, the results stated that the low HI in the welding zone leads to an inadequate softening of materials, which can result in issues such as porosity and shallow penetration. Conversely, excessive HI leads to the buildup of burrs at the edges of the welding bead and a reduction in the welding cross-section. Consequently, both insufficient and excessive HI can lead to a decrease in tensile strength. And, microstructural analyses reveal that the WZ (stir zone), exhibited significantly fine grain sizes and the grains in the WZ are of equiaxed shape, with a gradual reduction in grain size observed in the HAZ [23].

2.1.2. Ferritic stainless steel (FSS)

Fusion welding is the most common method used in the industrial setting to weld ferritic stainless steels (FSSs). In fusion welding applications, the choice of welding techniques and the associated parameters that go along with it are crucial for guaranteeing the formation of a strong joint. Several ongoing studies are currently exploring how welding parameters influence the mechanical, microstructural, and corrosion properties of FSS.

The study revealed that the various welding techniques, namely, GMAW, GTAW and shielded metal arc welding (SMAW), influence the mechanical and microstructural attributes of 409M FSS. Notably, among these methods, the GTAW process, with its reduced HI, results to a slower cooling rate and enhanced mechanical properties [24]. The utilization of GTAW with differing HIs (i.e., low, and high) impacts the microstructural, mechanical, and intergranular corrosion (IGC) features of 439 FSS [25]. In both low and high HI cases, the WZ showcases a dual-phase microstructure with ferrite grains and retained austenite (RA) situated at ferrite grain boundaries (GBs, see Figure 6.3 (a, b)). The RA volumetric fraction diminishes with increasing HI, accompanied by an expansion in the width of the HAZ. The prevalence of an increased RA volumetric fraction in low HI weldment constrains grain growth, thereby resulting in superior mechanical characteristics, like enhanced microhardness and tensile characteristics, when compared to its counterparts. Notably, the higher RA content near the ferrite GB in the low HI WZ causes greater chromium depletion close to the GB, consequently leading to a higher degree of sensitization (DOS) [25]. Utilizing cold metal transfer (CMT) welding with different HIs introduces changes in the microstructural, mechanical, and IGC characteristics of 439 FSS [26]. The solidification of both welds occurs via the FA mode, leading to the creation of δ ferrite as the primary phase within the austenite matrix (Figure 6.3 (c, d)). The WZ of the low HI case comprises lathy δ ferrite, whereas the WZ of the high HI case features skeletal δ ferrite, primarily attributed to differences in cooling rate, with the volumetric fraction of δ ferrite decreasing with increasing HI. In both weldments HAZ, coarser grains are observed near the FB due to the slower cooling rate, while finer grains are formed further from the FB, and the width of the HAZ increases with higher HI (Figure 6.3 (e, f)). The superior mechanical characteristics, including increased microhardness and tensile properties, observed in the low HI weldment are attributed to the higher δ ferrite content. IGC resistance, measured through DOS, is higher for the low HI WZ due to the prevalence of a higher δ ferrite content, which is considered a more active phase for sensitization [26].

The effect of HI on microstructural and mechanical characteristics of 430 FSS welded through the CMT method has been documented [27]. In each sample, the WZ exhibits a dual-phase microstructure comprising both ferrite and austenite. Additionally, the HAZ is subdivided into a coarse grain HAZ (CGHAZ) and a fine grain HAZ (FGHAZ). Notably, carbide precipitation is observed in the ferrite grains, while martensite formation occurs at the boundaries of the ferrite grains. In contrast to the fine grain zone, which features fine ferrite grains with minimal intergranular martensite, the coarse grain zone contains larger ferrite grains and a notable presence of intragranular carbides and intergranular martensite. Remarkably, the tensile fractures in all cases originate from the BM, indicating that the WZ exhibits superior strength compared to the BM [27]. Furthermore, the HI has a pronounced influence on the microstructural and mechanical characteristics of 409 FSS welded through the GMAW process [28]. The joints were prepared using two distinct austenitic filler electrodes (i.e., 304L and 308L) and three different HI values (0.3, 0.4, and 0.5 KJ/mm). The percentage dilution (%D) was calculated for all the weldments, revealing an initial increase and subsequent decrease with HI. The maximum %D occurs with

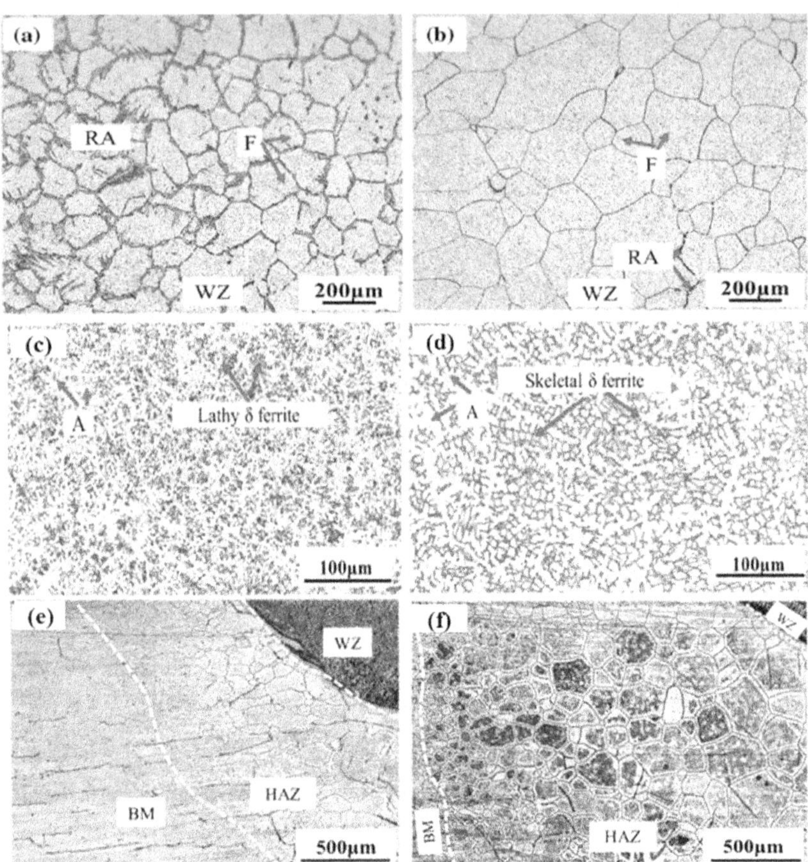

FIGURE 6.3 Optical microstructure of WZ of 439 FSS of (a) low HI (b) high HI [25]; WZ of CMT welding of (c) low HI (d) high HI; HAZ of CMT welding of (e) low HI (f) high HI. HI Note: HI- heat input [26]

medium HI, primarily due to the employment of the highest current. %D is chiefly affected by the distribution of heat between the BM and filler metal, which is primarily dictated by welding parameters like current and speed. The microstructure of the WZ in all joints comprises a mixed phase, including austenite, ferrite, and martensite. The type of phase formed in the WZ depends on the ratio of chromium to nickel equivalent (Cr_{eq}/Ni_{eq}). When this ratio surpasses 1.35, the initial phase formed during cooling is ferrite, and as cooling progresses, a portion of the ferrite phase transforms into austenite at varying temperatures, contingent on the Cr_{eq}/Ni_{eq} ratio. A higher Cr_{eq}/Ni_{eq} ratio, observed in the medium HI case, leads to a greater formation of martensite, which is a harder phase contributing to the increased hardness of medium HI compared to the other weldments. Moreover, the higher martensite content hinders grain growth, resulting in finer grains in the medium HI case. This reduction in grain size in medium HI weldment is ascribed to the enhancement of tensile properties, as compared to other weldments [28].

Beyond the parameters linked to fusion welding methods, those distinctive to solid-state welding methods, such as FSW, especially the welding speed, significantly affect the microstructure and mechanical properties of the joints. This influence is largely due to the development of an extremely fine-grained equiaxed microstructure within the friction stir-welded joint. As a result, the mechanical properties of the joint are maintained at a level considerably higher than that of the BM [29]. The lower welding speed in FSW of 409 FSS resulted in root sticking which occurs when the hot weld metal remains in contact with the steel backing plate for an extended period, leading to thermal damage to the welding tool and higher welding speed resulted in groove defect and the FSW joints achieved highest tensile strength and impact toughness, reaching 574 MPa and 37 J, respectively, when utilizing the optimized parameters: a rotational speed of 1460 rpm, a welding speed of 40 mm/min, and a shoulder diameter of 20 mm [30].

2.1.3. Duplex stainless steel (DSS)

The HI has a notable influence on the microstructural and corrosion characteristics of UNS S32750 super duplex stainless steel (DSS) welded using the SMAW process [31]. The E2595 electrode was employed with two different HI values, that is, 0.54 and 1.10 KJ/mm. In both weldments, various types of austenite, including grain boundary austenite (GBA), Widmanstatten austenite (WA), and intergranular austenite (IGA) are observed in the WZ, and the grain size in the WZ increases with rising HI. The development of various forms of austenite relies on the rate at which the cooling occurs; during cooling, GB austenite initiates the formation of GB δ ferrite, from which WA nucleates and grows within the grains. Apart from GBA and WA, if there is enough time for diffusion, IGA can also nucleate and grow. The HAZ in both weldments is notably narrow and devoid of any intermetallic precipitates. The tensile strength of both welded samples closely approximates that of the BM. Fracture in both cases occurs in a ductile manner. Furthermore, the hardness of the low HI weldment surpasses that of the high HI weldment. This disparity in hardness can primarily be attributed to variations in microstructural morphology and the distribution of micro-constituents within the WZ. The smaller dendrite size and a higher content of δ ferrite in the low HI WZ also contribute to the elevated hardness. Sensitization, as determined by the degree of Sensitization (DOS), demonstrates only slight differences between the two HI weldments because the width of the HAZ is nearly identical for both. Pitting resistance, determined via potentiodynamic polarization (PDP), reveals that the BM boasts a higher pitting potential than both weldments due to its balanced 50:50 ratio of ferrite and austenite phases. The slight divergence in the pitting potential of the two HI weldments is predominantly attributed to variations in the austenite-ferrite ratio [31].

Furthermore, the HI significantly influences the microstructure, mechanical characteristics, and corrosion resistance of welded joints in UNS 32101 lean DSS [2]. Welded joints were prepared using an ER2209 filler electrode, and three distinct HIs were employed (0.98, 1.05, and 120 KJ/mm). With increasing HI, the width of the WZ expands, and the microstructure includes δ ferrite and different forms of austenite (like GBA, WA, and IGA), aligning with prior literature [30]. The fraction of austenite increases with rising HI, contingent on the transformation time. In cases of low HI, the faster cooling rate allows insufficient time for $\delta \rightarrow \gamma$ transformation. The grain structure in the HAZ varies with HI when compared to the BM. These microstructural variations primarily result from the welding parameters, chemical composition, and groove geometry. However, no significant change in the width of the HAZ was observed from low to high HI. The hardness of the WZ marginally decreases with increasing HI, mainly attributed to differences in cooling and the austenite-ferrite ratio. Pitting corrosion of weldments was evaluated through PDP testing, revealing a decrease in pitting corrosion resistance with increasing HI. This decrease is primarily associated with the higher formation of secondary austenite, characterized by a lower concentration of Cr, Mo, and N. As HI increases, the pit density, size, and depth of pits on the surface also increase. Electrochemical impedance spectroscopy (EIS) results indicate that the diameter of the capacitance semicircle decreases with increasing HI, aligning with the PDP findings [2]. The HI likewise has an impact on the microstructure, hardness, and resistance to pitting corrosion in keyhole-TIG welded 2205 DSS [16]. The microstructure of WZs for all HI values consists of different types of austenite, including IGA, WA, and GBA as shown in Figure 6.4 (a–c). With increasing HI, the overall austenite content and WA content increase. In the WM, hardness rises with increasing HI. Within the WZ, the hardness of austenite exceeds that of the ferrite phase due to a higher content of solid solution nitrogen within the austenite, enhancing solid solution strengthening.

Additionally, the presence of a higher dislocation density further contributes to increased hardness. In each WZ, the ferrite phase exhibits a lower pitting resistance equivalent number (PREN) than the austenite. While there are no notable differences in the PREN of each WM, the corrosion rate in the WZ gradually rises as the HI increases. The formation of an intensive, stable, and thicker passive film by fine-grained IGA with a higher PREN may inhibit the formation and growth of pits. Consequently, the ferrite phase is selectively dissolved with a lower PREN [16].

In addition to the parameters associated with fusion welding techniques, the parameters specific to solid-state welding, like FSW, particularly welding speed, have an impact on the microstructural and mechanical properties of DSSs. In FSW, increasing the welding speed initially yields sound joints. However, beyond a certain threshold, groove-like defects start to form due to the reduced HI. The increase in welding speed, which effectively lowers HI, reduces the grain size of both austenite

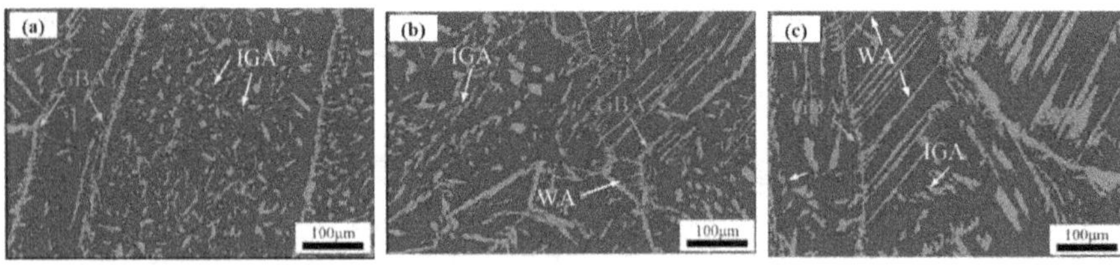

FIGURE 6.4 EBSD phase map of WZ 2205 DSS of different HI weldments (a) HI- 1.57 KJ/mm (b) HI- 1.74 KJ/mm (c) HI- 1.94 KJ/mm. Note: blue color represents the ferrite phase, and the red color represents the austenite phase [16]

and ferrite in the stir zone (SZ), resulting in an elevation of the tensile strength and hardness of the welded joints [32]. When joining DSSs, the choice of welding technique, whether it's the fusion method like GTAW or the solid-state approach like FSW, along with their specific parameters, including welding temperature, joint groove design, and the use of filler rods, significantly influences the microstructure. In GTAW, which involves a high HI and the use of filler rods, distinct phase morphologies, including GBA, IGA, and WA, are observed alongside the ferrite phase in the WZ. On the contrary, FSW welds exhibit a simpler dual-phase microstructure consisting of both austenite and ferrite, characterized by a marked grain refinement. The ferrite/austenite ratio in the WZ is influenced by the welding techniques, whether they belong to solid-state welding or fusion welding categories, as well as their specific parameters. This includes welding temperature, joint groove design, and the use of filler rods. The grain refinement observed in the SZ of FSW results in a substantial improvement in hardness and tensile properties for the weld joint when compared to those produced by GTAW [33].

2.1.4. Martensitic stainless steel (MSS)

The HI exerts a significant influence on the microstructural and mechanical properties of GTAW welded joints in lean super martensitic stainless steel (MSS) [34]. Various HIs were employed, that is., 7.97, 8.75, and 10.9 KJ/cm. It's observed that as HI increases, the size of the weld bead also increases. All joint's WZ microstructure comprises δ ferrite within the martensite matrix, and with increasing HI, the δ ferrite content in the WZ increases due to a higher solidification rate. As HI increases, there is a noticeable increase in grain coarsening near the FB, while simultaneously, the δ ferrite content decreases. The microstructure and the proportion of δ ferrite in the HAZ predominantly rely on the welding process's thermal cycles. In cases of high HI, the reduced cooling rate in the HAZ promotes the formation of austenite and the dissolution of the ferrite phase. With increasing HI, the tensile strength of the weldments decreases. This reduction can mainly be ascribed to the changes in the δ ferrite content [27]. The application of laser-arc hybrid welding with various HIs has a significant impact on the characteristics of 420 MSS, encompassing microstructure and toughness [35]. In hybrid welding, the WZ consists of two distinct zones: the laser zone and the arc zone. The microstructure in the arc zone comprises δ ferrite, austenite, and martensite, while the laser zone exhibits a similar microstructure but with a lower content of austenite. The arc zone exhibits higher Cr and Ni content because, during hybrid welding, the alloying elements from the filler wire tend to accumulate in the arc zone. The increased Ni content in the arc zone promotes the formation of austenite, which contributes to good toughness. The HAZ is divided into three regions: the CGHAZ, FGHAZ, and intercritically reheated HAZ (ICHAZ). The microstructure in the CGHAZ comprises abundant martensite, ferrite, and intra- as well as intergranular carbides, whereas the FGHAZ consists of blocky ferrite, carbides and finer martensite. An Erichsen cupping test employed to assess the microstructure and toughness of the CGHAZ, which aligns with the observation that increased HI is associated with improved toughness, relies on grain size and the amount of inter- and intragranular carbides within the CGHAZ. The CGHAZ is the weakest region of the

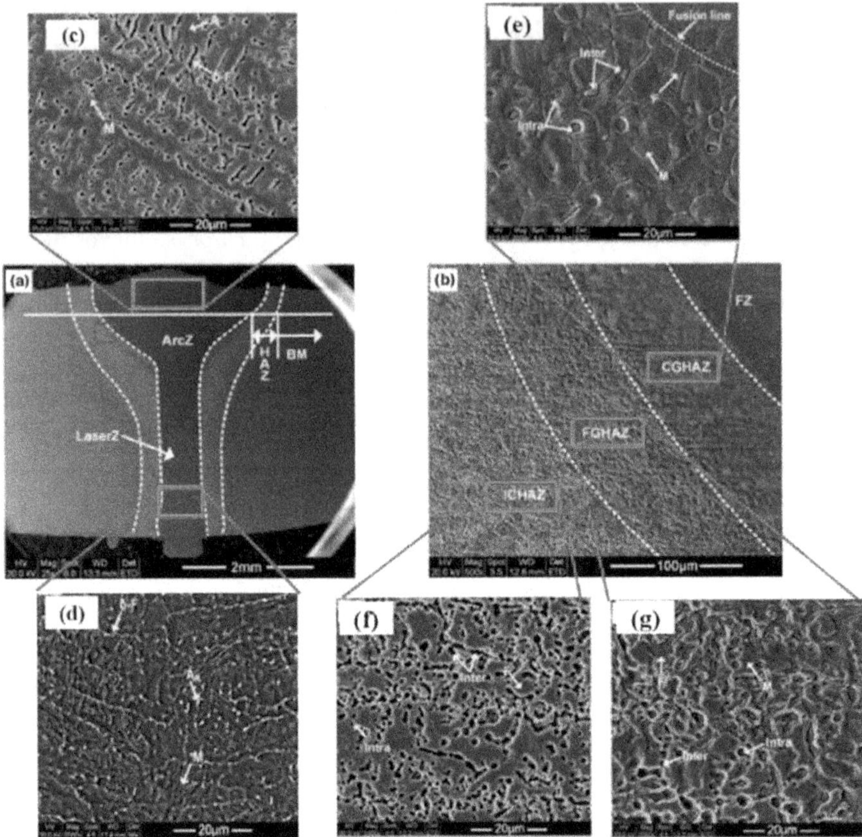

FIGURE 6.5 The microstructures of 420 MSS weld (a) weld profile, (b) HAZ; 420 MSS WZ microstructure, (c) arc zone, (d) laser zone; HAZ microstructure of weld of (e) CGHAZ, (f) ICHAZ, (g) FGHAZ [35]

weldments, with most joints fracturing from the CGHAZ, while some joints fracture from the base metal [35].

The weld geometry of AISI 420 MSS is greatly influenced by the welding parameters for pulsed Nd:YAG laser welding [36]. The parameters that affect the weld's dimensions are voltage, frequency, and laser diameter. The weld area, which includes the width and depth of the joints, increases as the voltage increases because a higher HI follows from this. The expansion of the weld width occurred due to a shift in the focusing point caused by an increase in laser beam diameter. However, when the average peak power density is reduced, it leads to a decrease in weld depth. The width and depth of the joint are both impacted by the pulse duration. The width and depth of the weld reach their ideal values as the pulse duration increases, which is brought on by an increase in pulse overlapping. However, the weld's depth and width decrease if the pulse duration goes beyond the optimal value because the average peak power decreases. The microstructure of the WZ contains limited retained austenite, martensite, and δ ferrite. Furthermore, the HAZ has the highest level of hardness because it contains $M_{23}C_6$ precipitates [36].

2.2. Dissimilar welding of SSs

Several challenges arise when fusing two different metals: pores form, elements are distributed unevenly, there is a chance that the joints may crack during solidification, and new phases may emerge [37]. Because HI plays a crucial part in the formation of strong joints, this section examines how HI affects the formation of sound dissimilar welds in stainless steels.

2.2.1. Austenitic-austenitic SS

As the primary alloying element in conventional ASSs, Ni is essential for maintaining the austenite phase's stability even at room temperature. These specific alloys—of which the 300 series is a well-known example—are known as chrome-nickel (Cr-Ni) ASSs. However, industries have been actively looking for alternatives due to the rising cost of Ni, which has resulted in the adoption of low-Ni or Ni-free ASSs. The austenite phase stability in these alternative SSs can be achieved by elements like manganese (Mn) and nitrogen (N) [38] by replacing nickel to a greater extent. Due to this, a new series known as low-Ni ASS or Cr-Mn-Ni-N ASS has emerged; these are referred to as the 200 series. These excellent combinations of strength, ductility, corrosion resistance, and cost-efficiency make these low-Ni ASSs especially attractive. The formation of sound joints is paramount, and one of the most important ways to avoid solidification cracks is by adding ferrite to the weld zone of ASSs weldments. A direct consequence of the solidification process, primarily determined by the welding process parameters, is the presence of ferrite in the weldments. Consequently, one of the most important strategies for avoiding these flaws is to regulate the volumetric fraction of ferrite [38]. Hence, it is crucial to identify the optimal welding process parameters that produce the ideal ferrite volumetric fraction, which will ultimately improve the component's corrosion and mechanical properties and increase its lifespan.

In dissimilar welding of the 300/200 series, the WZ typically consists of a mixed microstructure comprising both austenite and δ ferrite [38]. However, the volumetric fraction of δ ferrite varies depending on the welding process parameters. Moreover, different types of delta ferrite, namely lathy ferrite and skeletal or vermicular ferrite, form based on these process parameters, as depicted in Figure 6.6 [38]. The alteration in process parameters leads to variations in the cooling rate, which, in turn, impacts the microstructure. In ASSs, the solidification mode is generally FA, which implies that the initial microstructure during solidification is ferrite and gradually transforms into austenite as solidification progresses. The HI or welding speed influences the cooling rate during solidification. A slower cooling rate provides adequate time for ferrite transformation into austenite and vice versa. The different morphologies of δ ferrite result from restricted diffusion during the ferrite-austenite transformation at faster cooling rates [38, 39]. Furthermore, an increase in HI leads to an expansion of the HAZ on both sides of the BM. However, in the case of 201 ASS BM, the increase in HAZ is more pronounced compared to its counterpart, primarily due to its

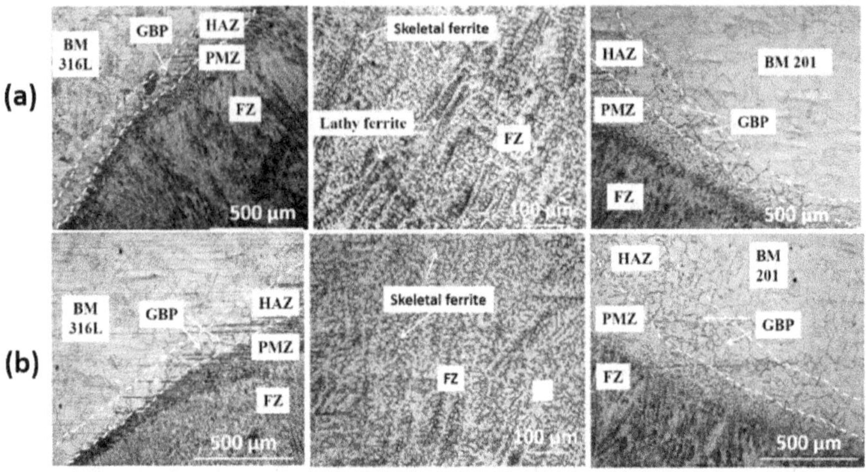

FIGURE 6.6 Optical microstructure of dissimilarly welded 316L-201 ASS by (a) low HI and (b) high HI process [38]

higher susceptibility to IGC [38]. Elevated HI also contributes to grain coarsening, increased dendritic length, and greater inter-dendritic spacing in the WZ, which is closely related to the mechanical characteristics of the weldments. Smaller dendritic length and reduced inter-dendritic spacing in low HI conditions result in improved tensile properties [38].

2.2.2. Austenitic-duplex SS

The dissimilar joining of ASS-DSS possesses many advantages in terms of cost and properties [40]. Notably, the DSS has replaced the use of ASS in applications where higher-stress corrosion cracking and pitting is required [41]. However, HI plays a vital role in retaining similar properties of the weldments.

The microstructures in dissimilar metal weldments can be divided into three different regions: the HAZ of the DSS, the WZ, and the unmixed zone (UMZ). As the HI increases, the width of the HAZ expands. Within the weld zone, as one moves from the center toward the HAZ, there is a gradual transformation in the microstructure, transitioning from fine dendritic grains to larger columnar grains. Moreover, isolated islands of austenite precipitate within the ferrite matrix. The amount of austenite present is affected by the thermal energy introduced during welding and the subsequent cooling rate [42]. The influence of HI on the evolution of microstructure in dissimilar SMAW of 2205 DSS and 316L ASS using a 2209 DSS filler electrode is depicted in Figure 6.7 [43]. Under high HI conditions, where cooling occurs more slowly, a greater portion of ferrite transforms into austenite compared to low HI conditions. The WZ showcases three distinct types of austenite: grain boundary allotriomorphs ($\gamma 1$), Widmanstatten austenite (γ WA), characterized by its acicular shape branching from allotriomorphs grains, and intergranular austenite ($\gamma 2$) located within the grains (Figure 6.7). Allotriomorphs grains and γ WA tend to nucleate at

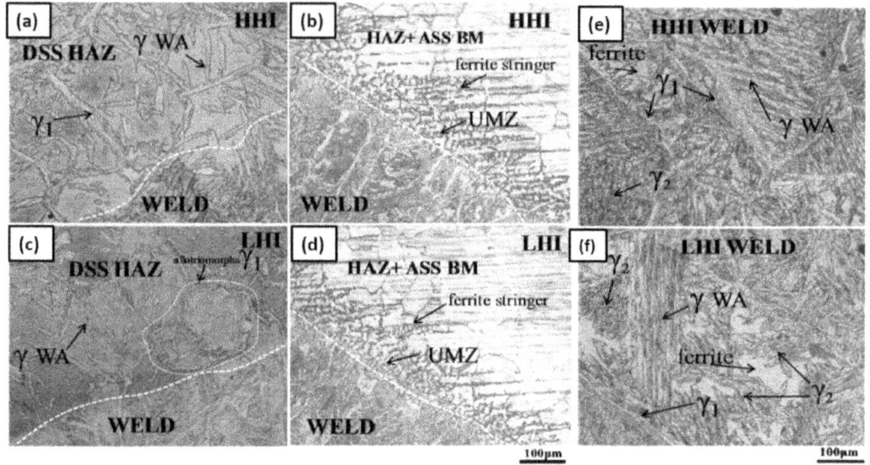

FIGURE 6.7 The microstructural evolution in dissimilar welded 316L ASS-2205 DSS by (a, b, e) high HI and (c, d, f) low HI process [43]

higher temperatures, while intergranular austenite forms only at lower temperatures with higher undercooling [43]. The growth of intergranular austenite necessitates lattice diffusion, which requires higher activation energy, making the process slower compared to the diffusion of allotriomorphs austenite. As a result, the growth of intergranular austenite is limited [44, 45]. Under lower HI conditions, slightly more γ2 austenite is observed than in higher HI conditions, mainly due to the faster cooling (Figure 6.7 (e, f)). The increased cooling rate promotes the formation of more secondary austenite in the weld [46].

2.2.3. Austenitic-ferritic SS

The joining of dissimilar stainless steels, including ASSs and FSSs, has become a widely adopted technique in various industrial sectors such as pressure vessels, heat exchangers, petrochemical, and so on [8].

In the case of dissimilar welding of ASS-FSS joints using GTAW technique, the WZ exhibits a dual-phase microstructure: ferrite and austenite [15]. The type of ferrite can vary and may include lathy, vermicular, or skeletal ferrite, depending on the HI applied during the welding process. When high HI and solidification within the FA range occur, vermicular, or skeletal ferrite formation takes place. This process involves the consumption of ferrite by austenite until the ferrite becomes depleted of elements that promote austenite (e.g., N and Ni) and enriched with elements favoring ferrite (e.g., Cr and Mo). Consequently, the ferrite stabilizes at lower temperatures, where diffusion is limited. On the other hand, low HI result in the formation of a lathy morphology instead of skeletal ferrite, because of limited diffusion during the transformation from ferrite to austenite (F→A) [18]. Near the FB on the ASS side, the UMZ forms, characterized by the presence of δ-ferrite stringers (Figure 6.8 (a, b)). On the FSS side, the HAZ can be separated into two clearly defined regions: the

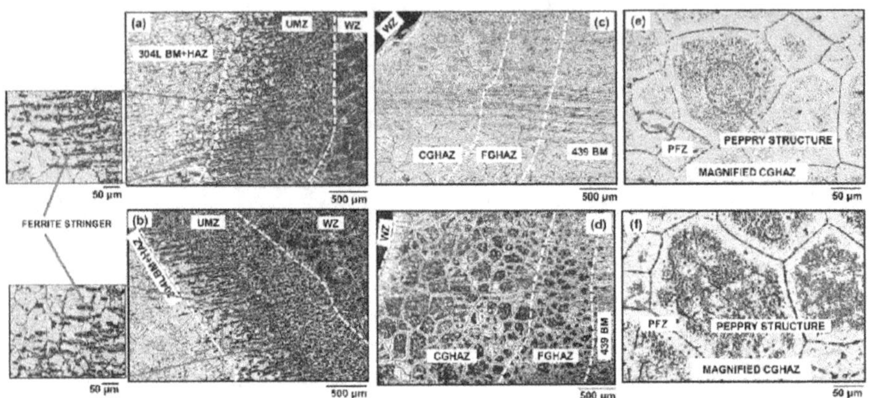

FIGURE 6.8 W1 (Low HI) weldment: (a) 304L ASS – UMZ, (c, e) 439 FSS - HAZ and magnified CGHAZ, and W2 (High HI) weldment: (b) 304L ASS – UMZ, (d, f) 439 FSS – HAZ and magnified CGHAZ [15]

CGHAZ and the FGHAZ (Figure 6.8 (c, d)). This division arises from differences in grain size resulting from the temperature gradients experienced during and after welding. Closer to the fusion boundary, higher temperatures and slower cooling rates lead to grain coarsening, giving rise to the CGHAZ. Conversely, further away from the fusion boundary, lower temperatures and faster cooling rates result in finer grains, forming the FGHAZ in that region [15, 47]. Within the HAZ region, precipitates, referred to as the "peppery structure" (Figure 6.8 (e, f)), are formed within the ferrite grains. This development arises from the diminished solubility of carbon in the ferrite phase at lower temperatures, creating a substantial force driving precipitation during the cooling process. In the vicinity of the grain boundaries, there exists a region referred to as the "precipitate-free zone" (PFZ), as depicted in Figure 6.8 (e, f). No precipitates are formed in this zone, as precipitation occurs at a specific distance away from the grain boundaries. The absence of these precipitates characterizes the PFZ, as they are found deeper within the ferrite grains, favored by specific conditions during the cooling process [18]. Moreover, the extent of the peppery structure's presence is noticed to be higher in the weldment with a HI. This is attributed to the increased dissolution of carbon within the ferrite matrix, leading to a higher number of supersaturation sites and subsequent precipitate formation.

2.2.4. Duplex-martensitic SS

The dissimilar welding of DSS and MSS is particularly valuable in crucial joints within the piping industry, where there is a simultaneous need for both corrosion resistance and high yield strength. In laser beam welding of 2205 DSS with 420 MSS, the rapid cooling in the MSS side during welding doesn't allow sufficient time for structural transformations, resulting in a microstructure primarily composed of dendritic martensite with a minor presence of δ-ferrite phases in the weld metal (WM) [48]. The initial microstructure in the MSS WM is δ-ferrite, which transforms entirely into austenite around 1100 °C. However, under lower-temperature cooling with an adequate cooling rate, some of the austenite transforms into the martensite phase, and some ferrite phase transformation may occur during the cooling process. If the formed ferrite phase contains ferrite-stabilizing elements, some retained δ-ferrite may remain along grain boundaries even after the transformation into austenite [49]. The rate of ferrite dissolution is influenced by the dissolution of ferrite-stabilizing elements, which results in the presence of some ferrite in the microstructure [49]. In the DSS side, the HAZ and WM mainly exhibit a ferritic microstructure and dendritic microstructure with coarse-grained austenite. Austenite formation can take place both on ferrite grain boundaries and within the grains themselves. The welding current significantly affects dissimilar welding of 2209 DSS and 410 MSS by upset resistance welding, and using an adequate welding current is crucial to achieving a sound joint [50]. A lack of adequate welding current may lead to the creation of gaps at the interface between the two BMs. In samples welded using lower welding currents, the existence of these flaws can be attributed to the lower temperature at the weld interface resulting from reduced welding current, which leads to lower heat generation. The WM on the MSS side typically exhibits a fully martensitic microstructure [51]. Furthermore, the microstructure on the side of the MSS is influenced by the carbon content in the BM. Specifically, in samples welded with lower welding currents (ranging from 2 to 3 kA), the microstructure consists predominantly of ferrite polygonal grains. Conversely, in samples welded with higher welding currents (3.5 and 4 kA), there is a potential for the formation of a small amount of non-equilibrium phase and martensite. The choice of welding current can impact the phase evolution. The Fe-Cr-C pseudo-binary equilibrium diagram for MSS [18] shows that temperatures below approximately 900 °C (corresponding to welding currents of 2 to 3 kA) do not lead to austenite formation. However, a very small amount of austenite may form at higher temperatures exceeding approximately 900 °C (corresponding to welding currents of 3.5 and 4 kA). Consequently, the microstructure may contain either minimal or only a slight amount of austenite. On the DSS side, the microstructure encompasses both ferrite and austenite phases, with the formation of different morphologies of austenite and ferrite depending on the welding current [50].

3. CONCLUSION

The impact of welding parameters on the welding of both similar and dissimilar SSs is examined, and the following conclusions can be inferred:

- The welding techniques and their parameters (such as welding current, voltage, and welding speed) significantly influence the weldments characteristics like microstructural, mechanical, and electrochemical.
- The various zones, i.e., WZ and HAZ forms during welding of SSs. However, the HI significantly impacts the width of HAZ. The higher HI usually results in the higher width of HAZ.
- The welding processes, filler and BMs chemical compositions primarily influence the solidification mode and consequently influences the microstructural evolution in WZ. The solidification mode can vary to F or FA or AF or A. In

SSs, based on the HI, different morphologies of ferrite or austenite can form in F and A mode, and, varying volumetric fractions of ferrite and austenite can form in AF and FA mode. For example, formation of dual microstructure, i.e. ferrite and austenite with varying volumetric fraction and different morphology of ferrite forms in similar and dissimilar welding of ASSs. Likewise, the various austenite forms found in the WZ of similarly welded DSS, including Widmanstatten austenite, grain boundary austenite, and intergranular austenite.

- Depending on the types of SSs, during welding, there can be formation of various zones in HAZ. For instance, in similar and dissimilar welding of FSS the HAZ is divided into the CGHAZ and FGHAZ.

REFERENCES

[1] C. Wang, Y. Yu, J. Yu, Y. Zhang, Y. Zhao, Q. Yuan. Microstructure evolution and corrosion behaviour of dissimilar 304/430 stainless steel welded joints. *J. Manuf. Process*. 2020;50:183–191.

[2] N. Ouali, K. Khenfer, B. Belkessa, J. Fajoui, B.C.B. Idir, S. Branchu. Effect of heat input on microstructure, residual stress, and corrosion resistance of UNS 32101 Lean Duplex stainless steel weld joints. *J. Mater. Eng. Perform*. 2019;28:4252–4264.

[3] A. Gupta, K.K. Verma, R. Kumar, C.S. Perugu, M. Prasad, S. Agrawal, et al. Electron backscatter diffraction analysis across the welded interface of post weld heat treated 2205 duplex stainless-steel and 316L austenitic stainless-steel dissimilar weldment. *J. Mater. Eng. Perform*. 2023;32:5493–513.

[4] G.M. Reddy, T. Mohandas, A.S. Rao, Y.V. Satyanarayana, Influence of welding processes on microstructure and mechanical properties of dissimilar austenitic ferritic stainless steel welds, *Mater. Manuf. Proces*. 2005;20(2):147–173.

[5] M. John, O. Diaz, A. Esparza, A. Fliegler, et al. Welding techniques for high entropy alloys: Processes, properties, characterization, and challenges. *Materials*. 2022;15.

[6] H. Choa, H.N. Hana, S. Hongb, J. Parkb, et al. Microstructural analysis of friction stir welded ferritic stainless steel. *Mater. Sci. Eng. A*. 2011;528:2889–2894.

[7] S. Saha, M. Mukherjee, T.K. Pal. Microstructure, texture, and mechanical property analysis of gas metal Arc welded AISI 304 austenitic stainless steel. *J. Mater. Eng. Perform*. 2015;24:1125–1139.

[8] J. Verma, R.V. Taiwade, R. Kataria, A. Kumar. Welding and electrochemical behavior of ferritic AISI 430 and austeno-ferritic UNS 32205 dissimilar welds. *J. Manuf. Process*. 2018;34:292–302.

[9] H. Agrawal, P. Sharma, P. Tiwari, R.V. Taiwade, R.K. Dayal. Evaluation of self-healing behaviour of AISI 304 stainless steel. *Trans. Indian Inst. Met*. 2015;68(4):501–511.

[10] E. Taban, E. Kaluc, A. Dhooge. Hybrid (plasma + gas tungsten arc) weldability of modified 12% Cr ferritic stainless steel. *Mater. Des*. 2009;30:4236–4242.

[11] H. Vashishtha, R.V. Taiwade, S. Sharma, A.P. Patil. Effect of welding processes on microstructural and mechanical properties of dissimilar weldments between conventional austenitic and high nitrogen austenitic stainless steels. *J. Manuf. Processes*. 2017;25:49–59.

[12] S. Bag, A. De. Development of efficient numerical heat transfer model coupled with genetic algorithm based optimisation for prediction of process variables in GTA spot welding Sci. *Technol. Weld. Join*. 2009;14 (4):333–345.

[13] P.K. Baghel. Effect of SMAW process parameters on similar and dissimilar metal welds: An overview. *Heliyon* 2022;8:12161.

[14] S.P. Tewari, A. Gupta, J. Prakash. Effect of welding parameters on the weldability of material. *Int. J. Eng. Sci. Technol.* 2010;2(4):512–516.

[15] S.K. Gupta, A.P. Patil, R.C. Rathod, V. Tandon, A. Gupta. Characterization of microstructure, mechanical and corrosion response in AISI 304L and TI-stabilized 439 stainless steels weld joints. *J. Manuf. Processes.* 2023;101:721–736.

[16] L. Han, T. Han, G. Chen, B. Wang, J. Sun, Y. Wang. Influence of heat input on microstructure, hardness and pitting corrosion of weld metal in duplex stainless steel welded by keyhole-TIG. *Mater. Charact.* 2021;175:111052.

[17] D. Shirmohammadi, M. Movahedi, M. Pouranvari. Resistance spot welding of martensitic stainless steel: Effect of initial base metal microstructure on weld microstructure and mechanical performance. *Mater. Sci. Eng. A.* 2017;703:154–161.

[18] J.C. Lippold, D.J. Kotecki. *Welding metallurgy and weldability of stainless steel.* A John Wiley and Sons; 2005.

[19] J. Yan, M. Gao, X. Zeng. Study on microstructure and mechanical properties of 304 stainless steel joints by TIG, laser and laser-TIG hybrid welding. *Optics Las. Eng.* 2010;48:512–517.

[20] S. Kumar, A.S. Shahi. Effect of heat input on the microstructure and mechanical properties of gas tungsten arc welded AISI 304 stainless steel joints. *Mater. Des.* 2011;32:3617–3623.

[21] A.V. Bansod, A.P. Patil, A.P. Moon, S. Shukla. Microstructural and electrochemical evaluation of fusion welded low-nickel and 304 SS at different heat input. *J. Mater. Eng. Perform.* 2017;26:5847–63.

[22] S.S. Kumar, N. Murugan, K.K. Ramachandran. Effect of friction stir welding on mechanical and microstructural properties of AISI 316L stainless steel butt joints. *Weld. World.* 2019:63:137–150.

[23] C. Meran, O.E. CanyurtFriction. Stir welding of austenitic stainless steels. *J. Arch. Mater. Manuf. Eng.* 2010;43.

[24] A.K. Lakshminarayanan, K. Shanmugam, V. Balasubramanian, Effect of welding processes on tensile and impact properties, Hardness and microstructure of AISI 409M ferritic stainless joints fabricated by duplex stainless steel filler metal. *J. Iron Steel Res. Int.* 2009;16:66–72.

[25] S.K. Gupta, A.P. Patil, R.C. Rathod, V. Tandon, H. Vashishtha. Investigation on impact of heat input on microstructural, mechanical, and intergranular corrosion properties of gas tungsten arc-welded Ti-Stabilized 439 ferritic stainless steel. *J. Mater. Eng. Perform.* 2022;31:4084–4097.

[26] S.K. Gupta, A.P. Patil, R.C. Rathod, V. Tandon, H. Vashishtha. Tailoring the process parameters for Ti-stabilized 439 ferritic stainless steel welds by cold metal transfer process. *J. Mater. Eng. Perform.* 2023;32:6042–6053.

[27] J. Zhou, J. Shen, S. Hu, G. Zhao, Q. Wang. Microstructure and mechanical properties of AISI 430 ferritic stainless steel joints fabricated by cold metal transfer welding, *Mater. Res. Exp.* 2019;6.

[28] S.K. Gupta, A.R. Raja, M. Vashista, M.Z.K. Yusufzai. Effect of heat input on microstructure and mechanical properties in gas metal arc welding of ferritic stainless steel. *Mater. Res. Express.* 2019:6.

[29] J. Han, H. Li, Z. Zhu, F. Barbaro, L. Jiang, H. Xu, Li Mac. Microstructure and mechanical properties of friction stir welded 18Cr–2Mo ferritic stainless steel thick plate. *Mater. Des.* 2014;63:238–246.

[30] A.K. Lakshminarayanan, V. Balasubramanian. Understanding the parameters controlling friction stir welding of AISI 409M ferritic stainless steel. *Met. Mater. Int.* 2011;17:969–981.

[31] A. Gupta, A. Kumar, T. Baskaran, S.B. Arya, R.K. Khatirkar. Effect of heat input on microstructure and corrosion behavior of duplex stainless steel shielded metal Arc welds. *Trans. Indian Inst. Met.* 2018;71(7):1595–1606.
[32] T. Saeid, A. Abdollah-zadeh, H. Assadi, F.M. Ghaini. Effect of friction stir welding speed on the microstructure and mechanical properties of a duplex stainless steel. *Mater. Sci. Eng. A.* 2008;496:262–268.
[33] M.M.Z. Ahmed, K.A. Abdelazem. Friction stir welding of 2205 Duplex stainless steel: Feasibility of Butt joint Groove filling in comparison to gas Tungsten Arc welding. *Materials*. 2021;14.
[34] C. Muthusamya, L. Karuppiahb, S. Paulraj, D. Kandasamid, R. Kandhasamy. Effect of heat input on mechanical and metallurgical properties of gas tungsten arc welded lean super martensitic stainless steel. *Mater. Research.* 2016;19(3):572–579.
[35] K. Hao, C. Zhang, X. Zeng, M. Gao. Effect of heat input on weld microstructure and toughness of laser-arc hybrid welding of martensitic stainless steel. *J. Mater. Proce. Technol.* 2017;245:7–14.
[36] S.H. Baghjari, S.A.A. Akbari Mousavi. Effects of pulsed Nd: YAG laser welding parameters and subsequent post-weld heat treatment on microstructure and hardness of AISI 420 stainless steel. *Mater. Des.* 2013;43:1–9.
[37] N. Arivazhagan, S. Singh, S. Prakash, G.M. Reddy. Investigation on AISI 304 austenitic stainless steel to AISI 4140 low alloy steel dissimilar joints by gas tungsten arc, electron beam and friction welding. *Mater. Des.* 2011;32:3036–3050.
[38] V. Tandon, M.A. Thombre, A.P. Patil, R.V. Taiwade, H. Vashishtha, Effect of heat input on the microstructural, mechanical, and corrosion properties of dissimilar weldment of conventional austenitic stainless steel and Low-Nickel stainless steel. *Metallogr. Microstruct. Anal.* 2020;9:668–77.
[39] W. Chuaiphana, L. Srijaroenpramong, Effect of welding speed on microstructures, mechanical propertiesand corrosion behavior of GTA-welded AISI 201 stainless steel sheets. *J. Mater. Process Technol.* 2014;214:402–408.
[40] B.P. Logan, A.I. Toumpis, A.M. Galloway, N.A. McPherson, S.J. Hambling. Dissimilar friction stir welding of duplex stainless steel to low alloy structural steel. *Sci. Technol. Weld. Join.* 2016;21:11–19.
[41] B. Deng, Y. Jiang, J. Gong, C. Zhong, J. Gao, J. Li. Critical pitting and repassivation temperatures for duplex stainless steel in chloride solutions. *Electrochim. Acta.* 2008;53:5220–5225.
[42] T. Chehuan, V. Dreilich, K.S. de Assis, F.V.V. de Sousa, O.R. Mattos. Influence of multi-pass pulsed gas metal arc welding on corrosion behaviour of a duplex stainless steel. *Corro. Sci.* 2014;86:268–74.
[43] J. Verma, R.V. Taiwade. Dissimilar welding behavior of 22% Cr series stainless steel with 316L and its corrosion resistance in modified aggressive environment. *J. Manuf. Process.* 2016;24:1–10.
[44] A.J. Ramirez, S.D. Brandi, J.C. Lippold. Secondary austenite and chromium nitride precipitation in simulated heat affected zones of duplex stainless steels. *Sci. Technol. Weld. Join.* 2004;9:301–13.
[45] A. Eghlimi, M. Shamanian, K. Raeissi. Effect of current type on microstructure and corrosion resistance of super duplex stainless steel claddings produced by the gas tungsten arc welding process. *Surf. Coat. Technol.* 2014;244:45–51.
[46] V. Muthupandi, P.B. Srinivasan, S.K. Seshadri, S. Sundaresan. Effect of weld metal chemistry and heat input on the structure and properties of duplex stainless steel welds. *Mater. Sci. Eng. A.* 2003;358:9–16.

[47] N. Kumar, A. Kumar, A. Gupta, A.D. Gaikwad, R.K. Khatirkar. Gas tungsten arc welding of 316L austenitic stainless steel with UNS S32205 duplex stainless steel. *Trans. Indian Inst. Metals*. 2018;71:361–72.
[48] C. Kose. Dissimilar laser beam welding of AISI 420 martensitic stainless steel to AISI 2205 Duplex stainless steel: Effect of post-weld heat treatment on microstructure and mechanical properties. *J. of Materi. Eng. and Perform*. 2021;30:7417–48.
[49] D. Carrouge, H.K.D.H. Bhadeshia, P. Woollin. Effect of δ-ferrite on impact properties of supermartensitic stainless steel heat affected zones. *Sci. Technol. Weld. Joi*. 2004;9:377–89.
[50] A. Ozlati, M. Movahedi. Effect of welding heat-input on tensile strength and fracture location in upset resistance weld of martensitic stainless steel to duplex stainless steel rods. *J. Manuf. Process*. 2018;35:517–25.
[51] M. Pouranvari, M. Alizadeh-sh, S.P.H. Marashi. Welding metallurgy of stainless steels during resistance spot welding Part I: Fusion zone and Part II: heat affected zone and mechanical performance. *Sci Technol. Weld. Join*. 2015;20:502–21.

7 Tungsten and Metal Inert Gas Welding: Automation in Welding

Virendra Pratap Singh, Basil Kuriachen, Vinyas Mahesh, and Dineshkumar Harursampath

1. INTRODUCTION

1.1. Tungsten inert gas (TIG) welding

TIG welding is a welding method that employs a non-consumable tungsten electrode to create a weld joint. This technique belongs to the category of arc welding, wherein the welding area is shielded from impurities in the atmosphere through the utilization of a protective gas, commonly helium (He). The utilization of He, an inert gas, in this context has led to the term "heliarc welding" in various sources. This procedure entails a consistent power supply to generate electricity, which subsequently move amid the arch through a sequence of ionized metal and gas, forming a plasma. This welding technique finds its primary application in working with non-ferrous metals like aluminium, copper, and magnesium.

Furthermore, it finds application in the welding of stainless-steel components. This welding technique offers a significant edge over SMAW and GMAW due to its capacity to regulate the welding procedure. As a result, this approach consistently delivers superior quality and more robust welds. Nevertheless, a constraint associated with this welding method is the intricate comprehension required for the process, coupled with a relatively slower operational pace compared to alternative welding processes [1].

A notable limitation involves the complexity of the manual TIG welding method, stemming from the precise focus demanded of the operator. Unlike many other welding techniques where the technician manipulates the filler material with one hand and manages the weld torch with the other, this technique necessitates the use of both hands [2]. As a result, ensuring the optimal arc length and preventing direct contact between the electrode and substrate becomes relatively intricate, leading to potential safety hazards. Additionally, initiating the weld arc necessitates a frequency creator to produce the necessary electric spark. This spark, directed through the shielding gas, initiates the arc and consequently separates the electrode from the workpiece. Following the arc's ignition, the operator passages the welding arc in a rotary manner

 DOI: 10.1201/9781003435884-7

to create a weld pool. Naturally, this process's success hinges on factors such as the required electrode current and the electrode's size. Moreover, the operator upholds a consistent gap between the electrode and substrate. Subsequently, the torch is stirred reverse, tilting within a vertical range of 10 to 25 °C. Here, additional filler material is put to end of weld pool as needed. Notably, filler rods are removed from the molten pool as the electrodes advance. Nevertheless, they must remain within the protective gas shield to prevent surface rusting caused by the welding process [3].

A crucial safety aspect to take into account while performing this welding procedure is the need for welders to maintain a sufficient distance from the arc, particularly when working with aluminium under a gas shield. This precaution is necessary due to the relatively low melting point of filler rods used for aluminium. If these rods come into very close proximity to the arc, there is a potential risk of them melting even before they make proper contact with the weld joint. Additionally, it is imperative to gradually decrease the arc current as the welding nears its conclusion. This gradual reduction allows for the proper solidification of the weld crater, thereby minimizing the likelihood of cracks forming at the conclusion of the welding process [4].

Ensuring safety is the utmost priority throughout this welding procedure, necessitating the utilization of safety gear. This gear includes delicate and lightweight leather gloves, as well as collared sleeves, designed to prevent any potential contact with hazardous ultraviolet rays. The reason behind this precaution is the absence of smoke amidst the process of welding, and the fact that the electric arc light lacks the protection provided by fumes, which is in contrast to shield metal arc welding. As a result, there is a heightened risk of operators being exposed to ultraviolet light.

The potential emission of arc light can lead to damage to both the eyes and the skin. Therefore, it is necessary to wear dark glasses in order to prevent exposure to ultraviolet rays. Furthermore, operators are susceptible to toxic smoke and particles released in welding operations due to the intense arc glare. These gases and particles contribute to the formation of nitric oxides in the ozone layer, giving rise to various health problems affecting the lungs and other parts of the body. As a result, welders may experience premature mortality [5].

Moreover, the heat released during welding can produce toxic fumes. One prominent sector employing TIG welding is the aerospace industry. This technique also finds application in various industries, particularly those involved in non-ferrous metal manufacturing. TIG welding is instrumental in crafting components for spacecraft and bicycles, as well as in the repair and maintenance of machine tools. The widespread adoption of TIG welding can be attributed to its notable attributes, including precise control of the weld area in comparison to alternative welding methods. This approach ensures the creation of high-quality welds that remain uncontaminated by substances like grease, and oil [5].

To ensure a high rated weld pool, it is essential that the flow of shielding gas remains sufficient and consistent. This consistency is crucial as it allows the gas to effectively envelop the weld, preventing any impurities from affecting it. This consideration is particularly important in TIG welding, where challenging conditions such as wind and dust can necessitate an increased gas shield for optimal weld protection. Consequently, this can lead to elevated process costs.

An eminent issue encountered in TIG welding is controlling the heat generation. The extent of heat, which can be either insufficient or excessive based on the current provided, plays a pivotal role in determining penetration depth and the positioning of the weld bead relative to the weld surface. Striking the right balance is crucial because overly high heat generation can cause the weld bead to expand in width and even result in the formation of droplets from the molten metal. Moreover, when the gap between the workpiece and welding torch is substantial, deficiencies in shielding gas coverage arise. This deficiency can give rise to weld porosity, characterized by the creation of pinholes in the weld. Consequently, welds afflicted by porosity are weaker compared to standard welds [4].

Furthermore, surpassing the advised current for the electrode can lead to the development of tungsten particulate in the weld, a phenomenon termed tungsten spitting. This concern can be addressed by modifying the electrode diameter. Another aspect impacting weld quality involves insufficient shielding of the electrode by the gas shield while in contact with the molten metal. This can result in both impurity and instability within the welding arc, necessitating extra expenses to grind the electrode and eliminate abrasions using emery papers [6].

1.2. Metal inert gas (MIG) welding

MIG, also referred to as GMAW, is a technique where an electric arc is generated between a wire electrode and the base metal. This arc produces the heat necessary to fuse the workpiece for joining. The shielding gas serves the purpose of safeguarding against contamination through the welding gun. This welding procedure entails maintaining a consistent voltage and supplying power through direct current. Within this welding process, the transfer of metal can be executed through four distinct manners: short-circuiting, pulsed-spray, spray, and globular methods. Each of these methods possesses distinctive properties, along with their own set of pros and cons, regardless of the intended use [6].

It has also been designed for the intention of welding non-ferrous material. In addition, when compared to other forms of welding, it exhibits a notably high welding speed when working with steel. Currently, best among the welding methods employed in industries, owing to its rapidity, compatibility with robotic applications, and versatility. Unlike TIG welding, it doesn't necessitate the utilization of shielding gas. Instead, it utilizes a hollow electrode wire that has been integrated with flux [7].

It finds its primary application in the fabrication of sheet metal field, also in the automotive industry. In scenarios like these, arc welding is predominantly employed to substitute resistance welding. It is also utilized extensively within automatic welding processes, for instance those involving robotics [6]. Nevertheless, the utilization of this technique outdoors presents challenges due to the dispersion of shielding gas by drafts, which leads to the infiltration of contaminants into the joint. To tackle this problem, arc welding is preferred for outside applications, particularly in the infrastructure sector. Moreover, unlike shielded metal arc welding, MIG welding is not commonly utilized for underwater welding. In MIG welding, the welding electrode takes the form as MIG wire, and the selection of this electrode depends on factors like the elemental composition of the metal being joined, variations in joint design, and the materials surface properties. Consequently, the choice of electrode significantly impacts the mechanical performances and overall quality of the welded material [8]. However, achieving a high-quality weld using this approach requires the weld to be uncontaminated, porosity, and defects. Hence, electrodes containing small amounts of Mn, Ti, Al, and Si are recommended to prevent porosity caused by the presence of oxygen. The size of the electrode must be in the range of 0.7 to 2.4 mm in diameter, and in some cases, even up to 4 mm, depending on variations in the process and the base metal being welded [7].

When it comes to quality, it's important to manage defects that arise due to the existence of porosity in the weld. These issues can lead to a lack of strength and malleability in the welded substance. To illustrate, dross is commonly linked with aluminium due to the occurrence of Al_2O_3 or N^{3-} ion in the substrate or electrode. Consequently, to prevent these challenges, it is advised to consistently brush and chemically treat the work piece and electrode to eliminate the oxide [8].

The primary reason for porosity is linked to the trapping of gas within the molten pool. This phenomenon takes place while the metal is in the process of solidifying, before the gas can escape. The source of this gas, however, is the impurities found in both the protective shield and the substrate [9]. The amount of trapped gas is directly tied to how quickly the weld pool cools down. It's worth highlighting that aluminium, due to its high thermal conductivity, is particularly susceptible to rapid cooling during welding. Consequently, this makes the occurrence of increased porosity more likely. Nonetheless, this situation can be managed by ensuring proper cleaning of the electrode and substrate during the fabrication. Additionally, the cooling rate can be regulated through preheating, which reduces the temperature difference between the welding area and the base metal [10].

By contrast, tungsten inert gas welding exhibits remarkable versatility. This leads industry experts to combine numerous materials effectively. Nevertheless, it operates at a slower speed when compared to MIG welding. As a consequence, this leads to longer lead times and elevated production costs. Furthermore, operators must undergo specialized training to attain the necessary precision and accuracy. Despite this requirement, the welding process can be finely controlled, yielding welds of both strength and precision. Generally, it delivers aesthetically pleasing welds [11].

Nonetheless, the metal inert gas welding procedure is typically utilized for thicker and larger materials, where the consumable wire serves the dual purpose of both electrode and filler material. When compared to the TIG method, it boasts significantly higher speed, leading to shorter production times and reduced costs. Moreover, it is also more user-friendly, easier to grasp, and results in welds requiring less post-weld cleaning in contrast to the TIG process. However, concerning the weld quality, the resultant weld is not as robust and immaculate as those achieved through the TIG process. Refer to Figures 7.1 and 7.2 for visual representations of the TIG and MIG processes.

In conclusion, welding aluminium proves to be notably more challenging when compared to welding steel. This is primarily because aluminium exhibits dissimilar behavior to steel. For instance, while steel readily reacts during the welding

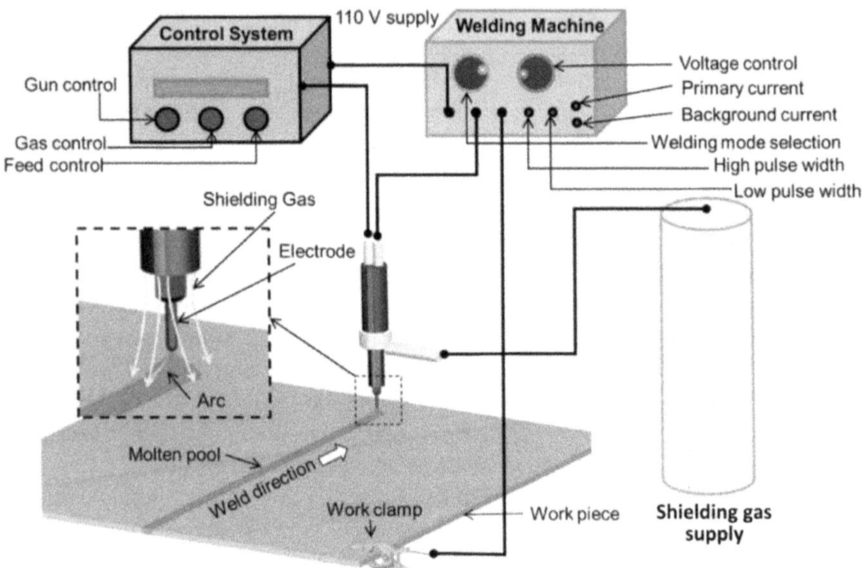

FIGURE 7.1 Schematic of the TIG process [12]

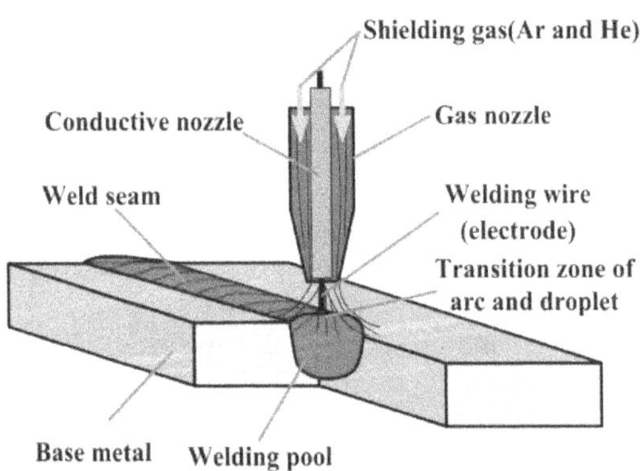

FIGURE 7.2 Schematic of the MIG welding process [13]

process, allowing for visual assessment, aluminium's welding necessitates reliance on judgment and experience [7]. The rationale behind this lies in the fact that aluminium develops an oxide layer on its surface, which requires thorough cleansing prior to welding. This oxide layer possesses a high melting point, around 2038 °C; in contrast, the actual Al beneath the oxide layer melts at a much lower temperature of 650 °C. Consequently, the removal of the oxide layer demands elevated heat, necessitating a cautious approach to prevent damaging the underlying aluminium. To address this predicament, the implementation of AC TIG (Alternating Current Tungsten Inert Gas) proves crucial, given its alteration between positive and negative currents in the electrode.

2. TIG WELDING FOR ALUMINIUM

Friction stir welding demonstrates suitability for joining aluminium due to its ability to mitigate solidification defects commonly encountered in traditional fusion welding methods. This welding technique yields impressive mechanical properties [7]. Nevertheless, it is linked to imperfections like micropores that manifest at the weld joint. This issue can be attributed to the lower heat supply inherent in FSW and aluminium's excellent heat conduction. Consequently, the insufficient heat input leads to plastic deformation, resulting in the formation of voids as defects. Of noteworthy consequence, these void defects serve as potential sites for crack initiation. Consequently, the eradication of such defects becomes crucial for enhancing the mechanical performance of joints produced via the FSW method [9, 14-17].

In pursuit of this objective, distinct joints consisting of 2.5-mm-thick Al-alloys were fabricated utilizing both FSW and TIG techniques. Tensile testing revealed an impressive 16.9% enhancement in tensile strength and a substantial 209.1% increase in elongation, both of which can be attributed to the implementation of the TIG arc-assisted approach. Furthermore, the outcomes indicated the successful elimination of the void defects typically observed in traditional fusion welding, leading to an enriched finer precipitate distribution. This advancement directly contributes to the enhancement of mechanical performances [18]. Magesh et al. [19] recognized that plastic deformation arises from combined friction and mechanical actions during the welding. Notably, FSW plays a pivotal role by virtue of its characteristics such as a refined microstructure, induction of residual stress in the weld, minimized porosity, and reduced incidence of weld cracks. Presently, the utilization of friction stir welding for aluminium welding is experiencing growing interest due to aluminium's widespread application in aerospace and automotive industries. Nevertheless, a significant challenge persists in the full integration of aluminium into car manufacturing, specifically concerning fusion issues, particularly in the lamination joining process [11]. Therefore, Magesh et al. [19] utilized an automatic technique for TIG in the context of Al welding. Their objective was to enhance the weld quality and strength, focusing on aspects like the dimension of weld penetration. The research incorporated specific control factors including arc length and weld speed. The GTAW process entails managing the tungsten electrode, electric arc, and the substrate. The investigation also delved into the mechanical attributes of the AA5059 alloy, covering tensile strength, fracture behavior, hardness, and microstructure. The findings indicated a reduction of approximately 30% in tensile strength during testing, with the base metal measuring 385 MPa and the welded portion at 268 MPa. Additionally, the study noted the presence of cracks at the weld points due to additional welding heat, as depicted in Figure 7.3. Overall, the research furnished valuable insights into the successful application of GTAW for the AA5059 alloy. It underscored the significance of input parameters for instance speed, current, and rate of gas flow in achieving improved mechanical performances [19].

The 2219 Al-Cu alloy has shown great promise for use in aerospace industries for manufacturing fuel tanks. This is primarily attributed to its remarkable strength and the ability to be welded effectively. The predominant welding method employed in producing rocket fuel tanks is the TIG welding process. This choice is driven by the

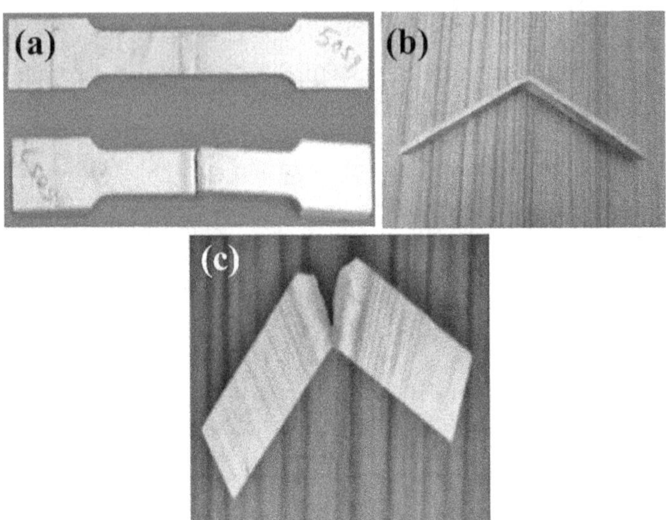

FIGURE 7.3 Graphical depiction of optimized samples [19]

weld quality it offers and its cost-effectiveness. Nevertheless, there are constraints associated with this technique. These limitations arise from the occurrence of tensile fracture initiation within the somewhat melted zone of the weld. Such initiation is a result of uneven solidification and stress concentration points within the welded area, as explained in reference [18].

It was documented that within the weld area, there exists a partially melted zone. This zone is distinguished by a slender region adjacent to the welding area where liquation occurs due to heating surpassing the eutectic temperature. Consequently, the semi-molten region is identified by eutectic grain boundary characteristics. Therefore, comprehending the composition of these eutectic structures becomes imperative, as they significantly influence the mechanical characteristics of the semi-molten area [10].

Constrained by the alloy's restrictions, Wan et al. [14] explored the less resilient section within the semi-molten area of the 2219-T8 alloy while employing TIG process. They achieved this by conducting a microstructural numerical simulation, aimed at assessing the tensile behavior of this semi-molten area. Moreover, the research delved into the inherent constitution and damage responses of both the eutectic structure and the aluminium matrix. Additionally, the investigation separately analyzed the stress-strain reactions leading to fracture, along with the tensile characteristics [20].

The findings demonstrated the presence of a predominant area with the highest concentration of eutectic particles at the weld joint, forming clusters. Moreover, these clusters considerably influence both elongation and tensile strength. Nevertheless, the plastic strain governs the behavior of the surrounding matrix, and notable stress accumulation occurs within the extensive eutectic structure, as noted [18]. Additionally, the occurrence of tensile cracks within the eutectic microstructure was observed; these initial cracks propagated into bigger fissures, ultimately leading to the eutectic microstructure fracture. Figure 7.4 illustrates the diverse characteristics of the Al-alloy while undergoing the fabrication technique.

Research conducted by Lu et al. [21] revealed that AA7055 alloy, created using metal spray forming process, can attain strengths of approximately 600 MPa post-deformation and heat treatment phase. Moreover, this Al-alloy possesses characteristics like ease of machinability and sufficient toughness. These exceptional attributes distinguish the alloy as a primary constituent in the railway and space technology.

Nonetheless, when the alloy is welded, it becomes prone to challenges such as stress-related corrosion and thermal fatigue fractures. Therefore, the identification and rectification of defects become imperative. Additionally, during the welding process of the spray-formed AA7055 alloy, multiple passes are required, resulting in the formation of a series of HAZ within the material [19].

Furthermore, it is commonly known that these HAZs constitute the weaker regions in the welded joints of the AA7055 alloy, and this weakness can be attributed to the potential occurrence of over-aging, causing softening. Similarly, the secondary thermal cycle typically affects the HAZ. Hence, it becomes crucial to monitor and comprehend the principles governing the microstructural changes and HAZ performance during the auxiliary heat loop to gain insights into the processes occurring within the HAZ.

In light of this, Lu et al. [21] conducted an investigation into the HAZ of various samples that underwent both the primary and a secondary heat transfer system. This simulation was conducted through the application of heat treatment procedures. The investigation primarily examined the impact of the softening effect brought about by the auxiliary heat loop and the HAZ dynamics during the TIG-welded deposition of AA7055 alloy using metal spray forming [21, 22].

The study found that the mechanical properties of the solid solution region deteriorated after the second thermal treatment. For instance, microhardness decreased from 180 to 103 HV, and tensile strength decreased from 570 to 346 MPa. These reductions were due to the weakening of the solid solution. However, higher treatment temperatures improved mechanical properties in the over-aging zone, ultimately leading to an expanded HAZ at the end of the auxiliary thermal treatment [23].

Temperature dispersal irregularity is linked to the AA2219-T8 alloy. Therefore, Wang et al. [24] conducted an inquiry into the mechanical behavior of two dissimilar Al-alloys, both of AA2219-T8. They achieved this by factoring in aspects such as residual stress, specimen size, and crack assessment, utilizing a FEM approach. The findings unveiled that the rate of residual stress release varied, notably decreasing with wider sample widths [24]. Furthermore, the outcomes of the tensile tests demonstrated a reduction in both tensile strength and elongation as specimen width increased. This discrepancy in tensile strength outcomes was attributed to size discrepancies rather than stress release, an insight gleaned from the Johnson-Cook crack assessment. Consequently, the investigation successfully formulated an alloy behavior prediction model employing this crack assessment [24].

In spite of the remarkable weldability exhibited by the AA2219 alloy, it experiences a decrease in strength both within the joint and across the broader welded area. This

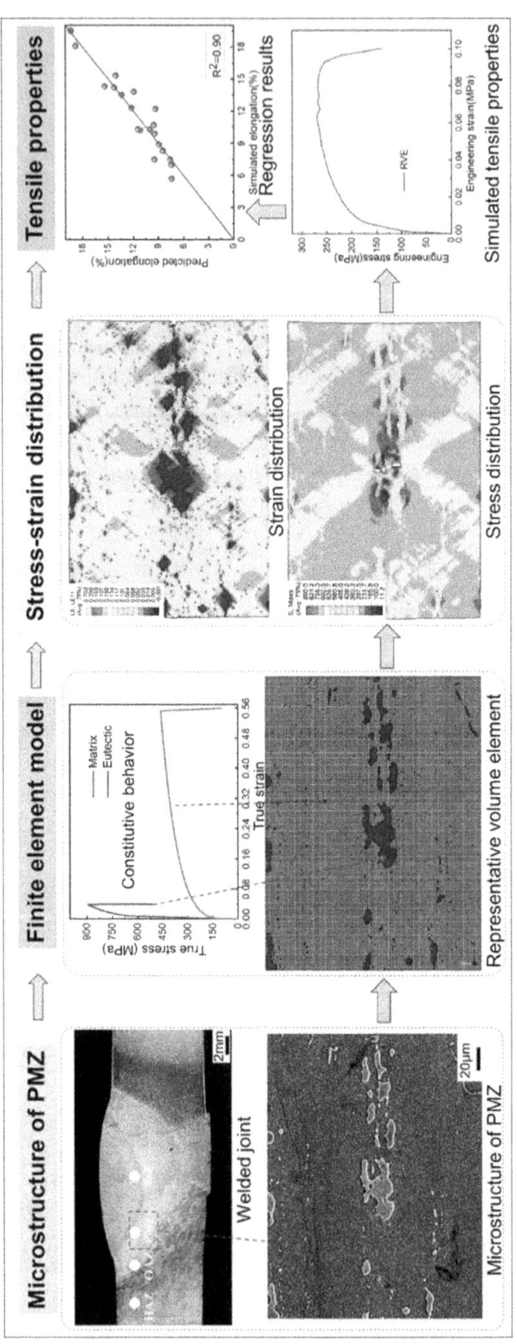

FIGURE 7.4 Characteristics of AA2219-T8 Alloy in TIG Welding [20]

consistent weakening at the joint is a result of strength decline. This reduction in strength at the joint is linked to the separation of elements within the alloy, the development of large grains during the welding process, and the dissolution of the intended strengthening precipitate. To tackle this issue, the utilization of weld reinforcements can be employed to enhance the joint's ability to bear loads [11].

Based on the suggestion provided, Wan et al. [23] employed the concept that utilizing the desired tensile attributes and robust joints in the AA2219-T8 alloy during TIG welding necessitates strategic control of the semi-molten zone and points of stress concentration. The outcome of their study indicated an enhancement in the quality of the semi-molten zone owing to the fine-grained structure and phase constituents present in the original material. Furthermore, there was a noticeable rise in the tensile strength, from around 16–21 MPa, which can be ascribed to the higher copper content and a reduction in the levels of coarse and continuous eutectic structures. In addition, there was a significant increase of about 318 MPa in the tensile strength. This was a consequence of the improvements made in the melting zone as well as the effective distribution of stress, as elaborated in a previous work [24].

Likewise, the application of AA7055 alloy in its sprayed form exhibits diminished strength and heightened brittleness because of the utilization of predictable welding wires during the fabrication process. These imperfections have a negative impact on its widespread utilization within the industry. Consequently, a study conducted by Huang et al. [25] delved into the exploration of two distinct types of aluminium alloy welding wires, both of which incorporated ceramic grains (TiB2 and ZrB2) into the Al-alloy spray form, and underwent TIG welding to create the weld joint. The outcomes indicated a prevailing microstructure characterized by TiB2-7055 grains with an equiaxed morphology. Furthermore, the resulting tensile strength reached approximately 51.6% of the base material, accompanied by an elongation of approximately 5%, showcasing a ductile fracture behavior. Additionally, in the case of the ZrB2-7055 joint, the microstructure attained a profusion of intricate dendritic formations. The attained tensile strength of 280 MPa represented around 46.6% of the substrate metal strength, indicative of a brittle fracture. These advancements contribute to mitigating the issues of softening and limited ductility in spray-formed AA7055. Another investigation by Lu et al. [22] explored the phenomenon of softening in the context of TIG welding of the metal spray-formed AA7055. The findings pointed to the softening occurring at the joint being associated with the dispersion of deposits in that zone.

3. MIG WELDING FOR ALUMINIUM

There is a growing need for lightweight, cost-effective components within the automotive sector. An appealing choice for this industry is Al-alloy due to its lightness and widespread accessibility relative to other metallic elements. Moreover, it exhibits favorable attributes such as ease of casting, remarkable strength, and resistance to corrosion. Nevertheless, to enhance its performance and application range, aluminium must be combined with other metals using various joining methods. An important issue arises from the excess hydrogen concentration in the die-casting of Al, stemming from rapid solidification in the casting process during aluminium production. This issue has significant repercussions, notably leading to porosity and thereby causing a reduction in the Al-alloy's mechanical performance [11, 23].

To tackle this concern, ultrasonic frequency pulses have been incorporated into the MIG welding procedure. The purpose of this integration is to effectively manage the occurrence of joint porosity by employing stirring and mixing techniques. This, in turn, results in an increase in the tensile strength of the welded joints. With the aim of achieving this goal, Ye et al. [19] adopted a method in which dissimilar aluminium alloy joints were joined together using the MIG process along with the application of ultrasonic frequency pulses. The results demonstrated a significant decrease in hydrogen porosity within the joints, leading to a substantial enhancement in the tensile strength, with an approximate increase of 185%. Consequently, the inclusion of ultrasonic frequency pulses into the process of metal inert gas welding proved successful in joining dissimilar aluminium alloys. This procedure yielded favorable microstructures and improvements in the tensile properties of the welded joints. Figure 7.5 illustrates visual representations of the welded joints created through the UFC-MIG technique, showcasing their morphologies and metallographic characteristics.

Moreover, a significant factor driving the preference for aluminium in the production of automobiles is its capacity to contribute to energy efficiency and reduced emissions. This advantage arises from its lightweight nature in comparison to steel, particularly noticeable in the fabrication of the vehicle's body [22]. Consequently, the exceptional and superior capabilities exhibited by aluminium alloys have led to a growing adoption of these materials in the automotive sector. Nonetheless, notable obstacles hinder the seamless integration of aluminium, encompassing issues such as disparate thermal characteristics among materials, limited solubility, and the formation of fragile compounds.

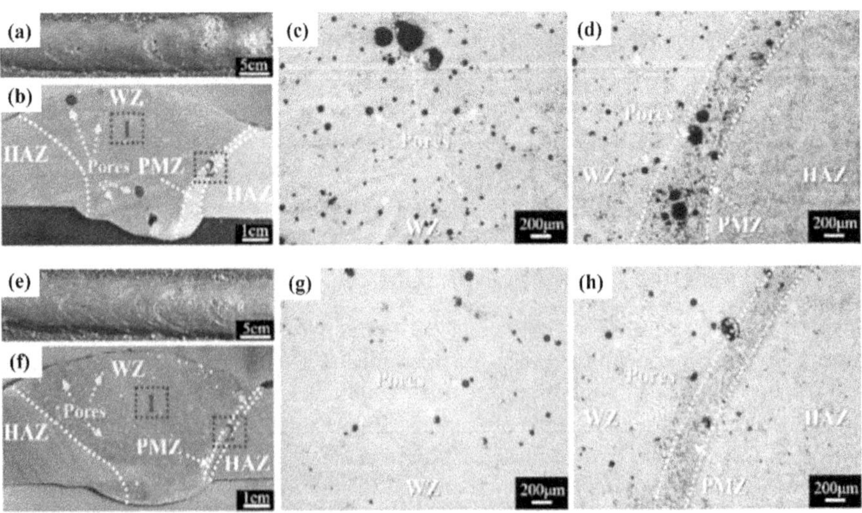

FIGURE 7.5 Microstructure image of UFC-MIG [26]

With the aim of achieving this objective, Wan et al. [20] introduced an innovative technique for MIG and brazing welding of Al and steel. This approach involved the integration of an external magnetic field to facilitate the movement of the arc and bead move in the linear direction, driven by an electromagnetic force across to the welding force. The outcomes demonstrated an enhancement in the spread of the molten alloy, attributed to the heightened current levels. Additionally, the weld joint's tensile strength was elevated due to the enhanced flowability of the molten metal due to EMF influence.

In the investigation by Arunakumara et al. [27], it was determined that the AA6000 series is utilized across multiple industries because of its remarkable ability to resist corrosion, ease of welding, and cost-effectiveness. Traditional fusion welding methods lead to issues like porosity, slag penetration, and incomplete integration of the Al-alloy. Consequently, the most suitable approach for fusion welding is the MIG welding process, as it guards the joint against impurities arising from welding. As a result, the study juxtaposed the evolutionary changes in fatigue and microstructural behavior of dissimilar Al-alloys, AA6082-T6 and AA6061-T6, when bonded using MIG and FSW methods [18, 28-30]. The investigation evaluated the tensile strength of both alloys, reliant on the chosen welding process settings. The outcomes indicated that the MIG method outperformed FSW, yielding improved results.

It is a well-established fact that the configuration of the weld seam plays a crucial role in influencing the strength, mechanical characteristics, and overall performance of aluminium alloys manufactured through metal inert gas welding, particularly in the contemporary field of fabrication. In line with this statement, Sharma et al. [31] anticipated the shape of the weld bead through precise optimization of model parameters, aiming to deliver exceptionally dependable welds.

A mathematical correlation was established to connect input parameters such as wire feed rate, speed, voltage, torch angle, penetration depth, weld width, and reinforcement height. The study employed statistical techniques including central composite design, variance analysis, and experimental design for analysis. It was noted that the mechanical properties and weld quality were influenced by the weld bead geometry [27]. The outcomes also suggested that the weld bead geometry is determined by the input parameters and can be optimized for the exquisite match. When it comes to joining Al and Mg alloys, which are frequently used in the automobile and aviation sectors due to their lightweight nature and ease of manipulation, the process is intricate. This is due to the formation of intermetallic compounds resulting from the reaction between Al and Mg, causing a decline in joint mechanical properties [31]. Although FSW and laser welding methods have been employed for alloy joining, these procedures are both intricate and expensive. Hence, the predicament revolves around controlling the creation of brittle intermetallic compounds during alloy joining. In addressing this, Zhang et al. [32] conducted MIG welding of 1 mm thick Al and Mg lap joints utilizing Zn foil as an interposing material. This interlayer served as a barrier to impede alloy reactions [33]. The results illustrated that the lap joint achieved was free from cracks, with the transitional layer comprising Mg and Zn. The tensile strength measured 64 MPa; however, fractures were identified at the fusion interface and the un-melted Mg-alloy region [34, 35-38].

4. AUTOMATION OF WELDING PROCESSES

Traditionally, the speed of the filler metal's delivery or the rate at which it's fed into the MIG welding process has been manageable at a consistent pace. Nevertheless, the manipulation of the torch is manual, signifying a semi-automatic procedure. Conversely, in the TIG welding process, both the torch and the filler metal are guided manually. Despite the welder's considerable skill, attaining uniform heat input throughout the welded seam proves to be highly challenging. Consequently, there's an uneven distribution of heat from one point to another, concluding the welding process. These divergent heat inputs lead to varying cooling rates and the subsequent development of distinct microstructures. Ultimately, this results in disparate mechanical properties along the welded path, an outcome incompatible with effective design approaches. As a solution, the automation of these welding machines is of utmost importance for both welding methods, ensuring consistent weld heat input by regulating parameters such as torch speed, filler metal feed rate, workpiece-to-nozzle distance, electrode-to-nozzle distance, and more. To realize this objective, specific gear train systems have been devised, deriving from the rotational speeds of the machine bed, enabling precise control over torch speed and filler metal feed rate.

In the context of TIG welding involving filler material, both TIG and MIG torches were affixed to the supported plate on a robotic machine, as depicted in Figure 7.6. The MIG torch in this arrangement was employed to automatically feed the filler metal and merge it with the primary TIG torch. By utilizing a wire feed motor, various speeds were chosen on the turning machine to achieve an optimal synchronization between the arc speed (associated with the TIG torch) and the rate of filler material delivery (linked to the MIG torch), thereby enabling the realization of completely automated TIG welding. Traditionally, in the MIG process, the feeding rate of the filler metal could be maintained at a constant pace, but the movement of the torch itself

FIGURE 7.6 Robotic advancement in TIG and MIG

required manual intervention, making it a semi-automatic procedure. Consequently, the challenges previously encountered in manual TIG welding resurfaced. However, this technique circumvents these issues by positioning the torch on the supported plate integrated into the turning machine bed, facilitating the transformation of the process into a fully automated welding operation.

5. SUMMARY

A comprehensive evaluation of the TIG and MIG welding methods has been conducted. The analysis revealed that both welding techniques are extensively utilized in industries such as automotive, aerospace, and food industries for the fabrication of Al-alloys. This preference is attributed to the advantageous properties of aluminium, including its lightweight, ease of casting, weldability, corrosion resistance and formability. Nonetheless, the assessment also uncovered a range of challenges faced by these welding methods. These include issues such as incomplete wetting, diminished strength, the formation of cracks at joints, porosity, and the development of fatigue and residual stress at the joints.

Furthermore, the study demonstrated that one notably effective approach to enhance the fatigue resistance of these welding processes is the utilization of FSW, particularly in the context of MIG welding. In either welding approach, a non-consumable tool is employed to penetrate the metal, generating frictional heat that induces plastic deformation. Consequently, this process leads to the creation of recrystallized microstructures, aligning with the aim to enhance the mechanical and microstructural characteristics. The integration of FSW with TIG/MIG welding has been demonstrated effectively in refining microstructures and enhancing mechanical characteristics. Both TIG and MIG welding techniques have demonstrated compatibility with automation, and automated versions of these processes yield favorable mechanical properties. The automation aspect contributes to a more consistent distribution of heat input along the welded seam.

REFERENCES

1. Ma, C., Chen, B., Meng, Z., Tan, C., Song, X., & Li, Y. (2023). Characteristic of keyhole, molten pool and microstructure of oscillating laser TIG hybrid welding. *Optics & Laser Technology*, 161, 109142.
2. Rajak, B., Kishore, K., & Mishra, V. (2023). Investigation of a novel TIG-spot welding vis-à-vis resistance spot welding of dual-phase 590 (DP 590) steel: Processing-microstructure-mechanical properties correlation. *Materials Chemistry and Physics*, 296, 127254.
3. Vaishnavan, S. S., Jayakumar, K., Kumar, P. N., & Suresh, T. (2023). Effect of ER5183 filler rod on the metallurgical and mechanical properties of TIG-welded AA5083 and AA5754 joints. *Materials Today: Proceedings*, 72, 2251–2254.
4. Ou, P., Cao, Z., Hai, M., Qiang, J., Wang, Y., Wang, J., ... & Zhang, J. (2023). Microstructure and mechanical properties of K-TIG welded dissimilar joints between TC4 and TA17 titanium alloys. *Materials Characterization,* 196, 112644.
5. Paturi, U. M. R., Vanga, D. G., Cheruku, S., Palakurthy, S. T., & Jha, N. K. (2023). Estimation of abrasive wear of nanostructured WC-10Co-4Cr TIG weld cladding using neural network and fuzzy logic approach. *Materials Today: Proceedings*, 78, 449–457.

6. Wu, M., Luo, Z., Li, Y., Liu, L., & Ao, S. (2022). Effect of heat source parameters on weld formation and defects of oscillating laser-TIG hybrid welding in horizontal position. *Journal of Manufacturing Processes*, 83, 512–521.
7. Wang, L., Ma, Y., Xu, J., & Chen, J. (2022). Improving spreadability of molten metal in MIG welding-brazing of aluminum to steel by external magnetic field. *Journal of Manufacturing Processes*, 81, 35–47.
8. Unni, A. K., & Vasudevan, M. (2023). Computational fluid dynamics simulation of hybrid laser-MIG welding of 316 LN stainless steel using hybrid heat source. *International Journal of Thermal Sciences*, 185, 108042.
9. Pan, D., Pan, Q., Yu, Q., Li, G., Liu, B., Deng, Y., & Liu, H. (2022). Microstructure and fatigue behavior of MIG-welded joints of 6005A aluminum alloy with trace amounts of scandium. *Materials Characterization*, 194, 112482.
10. Abima, C. S., Akinlabi, S. A., Madushele, N., & Akinlabi, E. T. (2022). Comparative study between TIG-MIG Hybrid, TIG and MIG welding of 1008 steel joints for enhanced structural integrity. *Scientific African*, 17, e01329.
11. Wu, D., Ishida, K., Tashiro, S., Nomura, K., Hua, X., Ma, N., & Tanaka, M. (2023). Dynamic keyhole behaviors and element mixing in paraxial hybrid plasma-MIG welding with a gap. *International Journal of Heat and Mass Transfer*, 200, 123551.
12. Khan, F. N., Junaid, M., Hassan, A. A., & Baig, M. N. (2021). Effect of pulsation in TIG welding on the microstructure, residual stresses, tensile and impact properties of Ti-5Al-2.5 Sn alloy. *Proceedings of the Institution of Mechanical Engineers, Part E: Journal of Process Mechanical Engineering*, 235(2), 361–370.
13. Li, C., Chen, Z., Gao, H., Zhang, D., & Han, X. (2021). Numerical simulation of the metal inert gas welding process that considers grain heterogeneity. *Proceedings of the Institution of Mechanical Engineers, Part L: Journal of Materials: Design and Applications*, 235(1), 42–58.
14. Singh, V. P., Patel, S. K., Kumar, N., & Kuriachen, B. (2019). Parametric effect on dissimilar friction stir welded steel-magnesium alloys joints: A review. *Science and Technology of Welding and Joining*, 24(8), 653–684.
15. Singh, V. P., Patel, S. K., & Kuriachen, B. (2021). Mechanical and microstructural properties evolutions of various alloys welded through cooling assisted friction-stir welding: A review. *Intermetallics*, 133, 107122.
16. Singh, V. P., & Kuriachen, B. (2022). Experimental investigations into the mechanical and metallurgical characteristics of friction stir welded AZ31 magnesium alloy. *Journal of Materials Engineering and Performance*, 31(12), 9812–9828.
17. Ranjole, C., Singh, V. P., Kuriachen, B., & Vineesh, K. P. (2022). Numerical prediction and experimental investigation of temperature, residual stress and mechanical properties of dissimilar friction-stir welded AA5083 and AZ31 alloys. *Arabian Journal for Science and Engineering*, 47(12), 16103–16115.
18. Yi, T., Liu, S., Fang, C., & Jiang, G. (2022). Eliminating hole defects and improving microstructure and mechanical properties of friction stir welded joint of 2519 aluminum alloy via TIG arc. *Journal of Materials Processing Technology*, 310, 117773.
19. Magesh, M., Arul, K., Subburaj, M., Rangaraja, R., Aravindh, B., Kumar, S. L., & Subbiah, R. (2022). Studies on mechanical and morphological of TIG welded aluminum alloy. *Materials Today: Proceedings*, 59, 1533–1536.
20. Wan, Z., Zhao, Y., Wang, Q., Zhao, T., Li, Q., Shan, J., & Wang, G. (2022). Microstructure-based modeling of the PMZ mechanical properties in 2219-T8 aluminum alloy TIG welding joint. *Materials & Design*, 223, 111133.

21. Lu, Z., Xu, J., Yu, L., Zhang, H., & Jiang, Y. (2022). Studies on softening behavior and mechanism of heat-affected zone of spray formed 7055 aluminum alloy under TIG welding. *Journal of Materials Research and Technology*, 18, 1180–1190.
22. Lu, Z., Yu, L., Xu, J., Du, C., & Zhang, H. (2022). Influence of secondary thermal cycle on softening behavior and mechanism of heat affected zone in TIG-welded spray formed 7055 aluminum alloy. *Journal of Materials Research and Technology*, 21, 2118–2132.
23. Wan, Z., Zhao, Y., Wang, Q., Zhao, T., Li, Q., Shan, J., … & Wang, G. (2022). Microstructure-based modeling of the PMZ mechanical properties in 2219-T8 aluminum alloy TIG welding joint. *Materials & Design*, 223, 111133.
24. Wang, Q., Wan, Z., Zhao, T., Zhao, Y., Yan, D., Wang, G., & Wu, A. (2022). Tensile properties of TIG welded 2219-T8 aluminum alloy joints in consideration of residual stress releasing and specimen size. *Journal of Materials Research and Technology,* 18, 1502–1520.
25. Huang, T., Xu, J., Yu, L., Cheng, Y., Hu, Y., & Zhang, H. (2021). Study on ductile fracture of unweldable spray formed 7055 aluminum alloy TIG welded joints with ceramic particles. *Materials Today Communications,* 29, 102835.
26. Ye, M., Wang, Z., Butt, H. A., Yang, M., Chen, H., Han, K., & Lei, Y. (2023). Enhancing the joint of dissimilar aluminum alloys through MIG welding approach assisted by ultrasonic frequency pulse. *Materials Letters,* 330, 133289.
27. Arunakumara, P. C., Sagar, H. N., Gautam, B., George, R., & Rajeesh, S. (2023). A review study on fatigue behavior of aluminum 6061 T-6 and 6082 T-6 alloys welded by MIG and FS welding methods. *Materials Today: Proceedings*, 74, 293–301.
28. Singh, V. P., Kumar, D., Mahto, R. P., & Kuriachen, B. (2023). Microstructural and mechanical behavior of friction-stir-welded AA6061-T6 and AZ31 alloys with improved electrochemical corrosion. *Journal of Materials Engineering and Performance*, 32(9), 4185–4204.
29. Singh, V. P., Patel, S. K., Kuriachen, B., & Suman, S. (2019). Investigation of general welding defects found during friction-stir welding (FSW) of aluminium and its alloys. In *Advances in Additive Manufacturing and Joining: Proceedings of AIMTDR 2018* (pp. 587–595). Singapore: Springer Singapore.
30. Singh, V. P., Kumar, R., Kumar, A., & Dewangan, A. K. (2023). *Automotive light weight multi-materials sheets joining through friction stir welding technique: An overview.* Materials Today: Proceedings.
31. Sharma, A., Verma, A., Vashisth, D., & Khanna, P. (2022). Prediction of bead geometry parameters in MIG welded aluminium alloy 8011 plates. *Materials Today: Proceedings*, 62, 2787–2793.
32. Zhang, H. T., & Song, J. Q. (2011). Microstructural evolution of aluminum/magnesium lap joints welded using MIG process with zinc foil as an interlayer. *Materials Letters*, 65(21–22), 3292–3294.
33. Huang, T., Xu, J., Yu, L., Cheng, Y., Hu, Y., & Zhang, H. (2021). Study on ductile fracture of unweldable spray formed 7055 aluminum alloy TIG welded joints with ceramic particles. *Materials Today Communications*, 29, 102835.
34. Wang, W., Yamane, S., Wang, Q., Shan, L., Zhang, X., Wei, Z., … & Xia, Y. (2023). Visual sensing and quality control in plasma MIG welding. *Journal of Manufacturing Processes*, 86, 163–176.
35. www.ipgphotonics.com/en
36. Venugopal, V., Singh, V. P., & Kuriachen, B. (2023). *Underwater friction stir welding of marine grade aluminium alloys: A review.* Materials Today: Proceedings.

37. Singh, V. P., Kumar, A., Kumar, R., Modi, A., Kumar, D., Mahesh, V., & Kuriachen, B. (2023). Effect of rotational speed on mechanical, microstructure, and residual stress behaviour of AA6061-T6 alloy joints through friction stir welding. *Journal of Materials Engineering and Performance*, 1–16.
38. Singh, V. P., Kumar, D., & Kuriachen, B. (2023). Effect of low welding and rotational speed on microstructure and mechanical behaviour of friction stir welded AZ31-AA6061-T6. *Transactions of the Indian Institute of Metals*, 1–9.

8 Friction Stir Welding and Design

Namdev Ashok Patil, Roshan Vijay Marode, and Sambhaji Kashinath Kusekar

1. INTRODUCTION

The most popular method for solid-state welding is friction stir welding. A subset of several welding processes called "solid-state welding" aims to weld materials below their melting temperatures without the need for fillers or brazing chemicals commonly employed in gas welding processes. For instance, involving challenging or apparently impossible material joining, friction stir welding (FSW) is very helpful. The Welding Institute developed it in 1991, and metal-matrix composites, magnesium alloys, titanium alloys, and aluminium alloys are among its frequent applications [1].

It has garnered increasing attention due to its ability to address prevalent challenges encountered in traditional fusion welding methods. Significant cracking during solidification, segregations, the occurrence of inherent porosities, the presence of solid inclusions, the development of intermetallics when interfacial reactions during dissimilar joining, high energy consumption, a larger heat affected region, and environmental pollution are just a few of the problems covered by these challenges [2].

Through the dissipation of heat generated by the friction between the rotating tool and workpiece, along with the application of plastic deformation, a solid-state weld is formed [3-6]. Because it can connect different materials of different thicknesses, it is cost-effective because it doesn't need costly consumables, and has the potential for full automation, FSW has become quite popular. This technique became popular from its initial application in aerospace metal components joining. Later it found extensive application across various industries, such as defense, railway, renewable energy, and the automobile sector [7]. The plunging revolution of a tool head with considerable force between two clamped metal parts is how the FSW process generates heat. Due to the enormous friction and heat generated, the metals start to circle around the tool. When the tool is traversed along the joint line, a piece of welded metal is then left behind. A groundbreaking solid-state welding method called FSW has evolved, revolutionizing the area of material joining [8]. With its capacity to produce high-quality joints in a variety of metal alloys and metal-to-nonmetallic material combinations, FSW has found use in a number of different sectors, including renewable energy, aerospace, automotive, defense, and the railway. This chapter explores the

DOI: 10.1201/9781003435884-8

most current developments in FSW technology, concentrating on how they improve design and performance.

The key principle behind FSW lies in the utilization of heat generated by the rotation and plunging of a specialized tool head with substantial force onto the interface of clamped metal parts. The resulting friction and heat cause the metals to undergo rotational movement around the tool, facilitating the consolidation and formation of a solid-state weld. This unique process eliminates the need for conventional melting and solidification, leading to improved joint integrity and reduced distortion [9].

To comprehend the intricacies of FSW, extensive research and analysis have been conducted, resulting in a wealth of knowledge and understanding of the technique. Various articles have been published, presenting detailed insights into the FSW process, its parameters, and its impact on different material combinations [10-17]. A family of alloys known as non-heat-treatable aluminium alloys is generally strengthened by solid-solution hardening or cold work. They are also referred to as work- or strain-hardened alloys. In addition, improvements in numerical simulation methods, including Finite Element Meshing Analysis, have arisen to support experimental studies and provide a practical and affordable method of improving FSW designs.

In this chapter, we aim to explore recent breakthroughs in FSW design and performance enhancements. We have delved into cutting-edge research on material selection, tool design, process optimization, and joint characterization. By examining these advancements, we can gain a comprehensive understanding of how FSW can further improve weld quality, strength, and overall performance in various applications.

Overall, this chapter aims to shed light on the latest developments in FSW technology, presenting the significance of design considerations and process improvements in achieving superior welds. By staying abreast of these advancements, engineers, researchers, and industry professionals can harness the full potential of FSW and drive innovation in the material joining.

2. FRICTION STIR WELDING DESIGN AND PARAMETERS

The schematic of the technique is indicated in Figure 8.1. The two metal plates to be joined are fixed on the back support plate fixture and clamped as shown in Figure 8.1. The clamping is very important as the high speed rotating FSW tool plunges at the joint of the plate and after initial dwelling it traverses along the joining line. The FSW tool pin is inserted in a controlled manner at one end, the pin fully submerged inside the tight metal plates joining line. Superplastic deformation occurs while the tool pin continuously blends the plasticized material from the advancing to retreating portion of the weld bead. The shoulder touches the plate upper surface and provides required pressure on the plasticized material to locally well consolidate it. Now the tool traverses with uniform speed horizontally and completes the welding. The process parameters to be carefully controlled for a defect-free weld are: tool rotation and linear speeds, tilt angle of tool, axial load on the FSW tool, tool plunge depth and initial dwell time. The FSW design also requires appropriate selection of FSW tool type, tool material, tool pin profile, and shoulder surface configurations. The dynamics

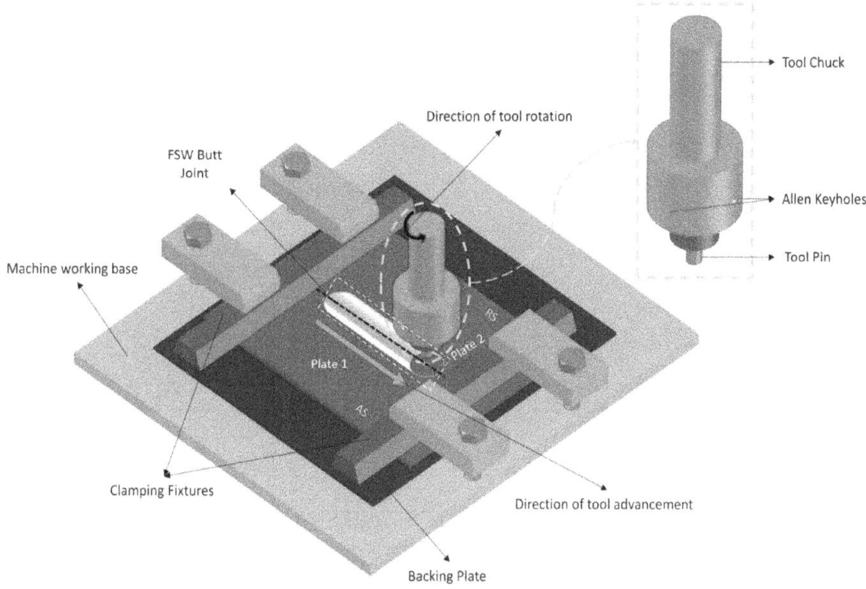

FIGURE 8.1 Configuration of friction stir welding

of the super plastic material movement depend on the tool design configurations. The different types of FSW tools are used in specialized FSW operations like refill FSW, non-weld thinning FSW, stationary shoulder FSW, and so on. Different tool pin profiles and shoulder profiles are used in these specialized FSW approaches.

There are different challenges for various approaches of friction stir welding and these are summarized in Figure 8.2. In order to overcome the challenges, the modified approaches are proposed and applied by researchers. Numerous studies have demonstrated that using inappropriate welding factors in FSW can result in the emergence of void defects in the joint; significantly compromising its mechanical properties [18-21]. The analysis of the FSW process through the thermal-fluid-structural coupling model has been conducted. It is observed that the plasticized material moves horizontally from the retreating side toward the rear end of the FSW tool and generally fills the space behind the tool continuously. However, due to non-uniform tool to workpiece contact pressure, the lower shear stress due to friction at the back side of the tool leads to substantial reduction in plasticized material flow velocity. Thus after passing retreating side, present plasticized material stagnates at the underside of weld; subsequently the void is formed in the middle and advancing side portion. Further the difference in the minimum and maximum contact pressure during FSW is must be less than 15 MPa in order to have sound weld bead [21]. In order to maintain higher tool-workpiece contact pressure, an increase in tool plunge depth is not advisable, as it leads to the reduction in the thickness of weld bead and the large continuous chip formation takes place [22]. It is also called a weld thinning effect. The weld thinning leads to deteriorated mechanical properties of the weld. High-quality FSW weld joints are required in the defense, aerospace, and automobile applications.

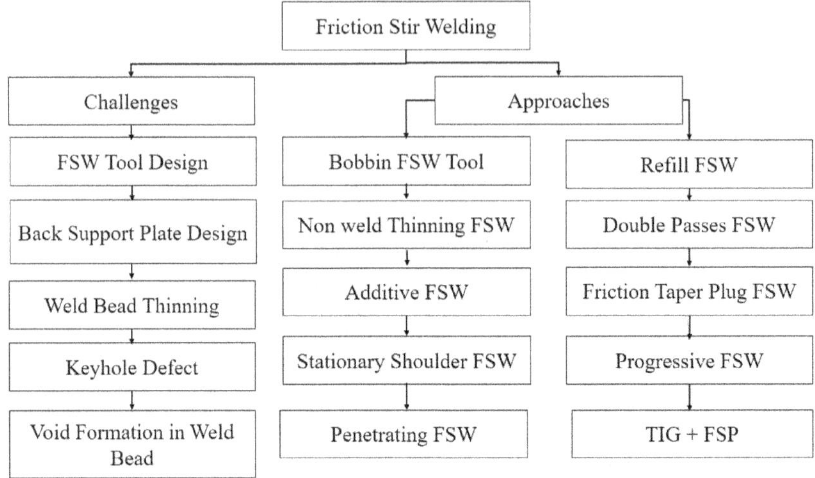

FIGURE 8.2 Challenges and approaches of FSW design

For overcoming the weld thinning effect, three approaches have been tried. The nonrotating shoulder which put the constant workpiece-tool pressure is utilized under the stationary shoulder FSW process. The second approach consists of the external additional material at the region of weld by means of an additive FSW technique. The third approach is of non-weld-thinning FSW, in which the non-plasticized material is taken out of weld bead by performing a flat weld process [23-25]. The stationary shoulder puts constant pressure and remains in contact with weld plates surfaces, wherein the rotating pin plunges and completes the process of joining.

The back plate supports the weld plates to be joined and is an important factor in FSW design. The rigid back plate support over which weld plates are clamped tightly avoids the possibility of throughout welding in the case of FSW. The welding of one of the base metal alloy plates with the support plate takes place in case of more plunge depth. So to overcome complexities of the fixture support back plate, the bobbin tool FSW was invented and successfully implemented in industrial applications. The bobbin tool FSW has two shoulders: lower and upper shoulder and between them the butt weld plates are processed by rotating pin. Here, the lower and upper shoulders act as the forging surfaces, which consolidates the plasticized material effectively, leading to sound quality weld bead [26-27].

The key hole defect at the end of the weld can be diminished using refill friction stir spot welding (FSSW). The FSSW tool has specialized design in which the clamping ring, sleeve and a pin are used in combination. The outer clamping ring does not rotate and the plunging is completed by rotating pin initially, and then by the sleeve [28-29]. The friction taper plug FSW process is used for the crack repairs for the under-water pipes, tank surfaces, and so on. The blind hole is drilled on the surface and then the plug material is inserted upon which the rotating pin plunges and space is filled with the plasticized plug and base metal material [30].

While there are several commercially accessible substitutes for FSW instruments, their application is restricted to extremely particular circumstances in contrast to traditional fusion welding consumables. The noteworthy points to take into account are that the pin's length is primarily determined by the thickness of the material to be welded, not by the aluminum series, and that the discrepancy between the two should be kept to a maximum of 5 to 6%. Comparing the 5XXX series to the 2XXX, 6XXX, and 7XXX series, lower shoulder and pin dimensions are needed. The shoulder diameters for the 2XXX and 7XXX series are comparable [31].

For the FSW of aluminum and other soft alloys, durable and reasonably priced tools are developed of different configurations. Although they are currently under development, they are required for the practical application of FSW to high-strength materials. The weld quality, tool wear, and tool performance are influenced by the qualities of the tool material, including strength, fracture toughness, hardness, thermal conductivity, and thermal expansion coefficient. Another crucial factor to take into account is the tool material's reactivity with both the workpiece and atmospheric oxygen. The tungsten (W)-based alloys and polycrystalline cubic boron nitride (PCBN) are significant contenders for the FSW of high-strength materials. When compared to other tools, the great strength, hardness, and high temperature stability of PCBN allow for far less wear [32].

The size and form of the tool shoulder and pin have an impact on the plastic flow and rate of heat generation in the workpiece. Weld characteristics, flaws, and tool forces are all impacted by tool design; yet, at the moment, empirical design is done through trial and error. The systematic creation of tools based on scientific ideas is still in its early stages of development. A few examples of recent research are the computation of tool shoulder dimensions based on the tool's grip of the plasticized material and the flow fields for various tool designs [32]. The material flow and heat generation rates are influenced by cross-sectional shape and surface characteristics, such as threads. In comparison to the tool shoulder, tool wear, distortion, and failure are also far more noticeable in the tool pin. It is possible to analyze the axial, longitudinal, and lateral forces on the tool based on actual data or to calculate them as functions of process parameters. It is necessary to estimate the tool pin's load-bearing capacity while taking into account the maximum stresses that result from the combined effects of bending and torsion. Coordinated research efforts are required to create long-lasting, reasonably priced instruments for the practical application of FSW to hard engineering alloys.

FSW design represents a critical facet of the welding process, encompassing a multitude of considerations to ensure optimal performance and weld quality. Central to this design is the FSW tool, characterized by a rotating shoulder and a non-consumable pin. The tool's geometry is a meticulous interplay of dimensions, with the shoulder and pin configurations tailored to induce plastic deformation in the workpiece material, facilitating a solid-state weld. For instance, in applications involving dissimilar materials, a specially designed tool with adaptive features may be employed to accommodate variations in thermal and mechanical properties [33-35].

Material selection for the FSW tool is a key factor influencing its performance. Considerations for high heat resistance and wear resistance are paramount to extend the tool's lifespan. Tungsten-based materials, for example, are often chosen for their robustness in handling the intense heat generated during the welding process.

Cooling mechanisms integrated into the tool design serve to dissipate the substantial heat generated during FSW. Efficient heat management not only ensures the longevity of the tool but also plays a crucial role in preventing thermal distortion in the workpiece. A prime example is the use of water-cooled FSW tools, where a flow of coolant mitigates temperature spikes and maintains the tool's structural integrity. Adaptive features in FSW tool design cater to the varied demands of different welding scenarios. For instance, a tool with an adjustable pin may be employed to accommodate varying workpiece thicknesses, contributing to the versatility of the welding process. Precision in alignment and positioning systems is fundamental to achieving accurate and consistent welds. Tool attachment mechanisms, such as quick-change systems, further enhance operational efficiency by facilitating easy tool replacement. The role of defects in FSW tool design is an important aspect to consider. Parameters like tool traverse speed, rotational speed, and axial force can influence the occurrence of defects such as tunneling, kissing bonds, or voids. Balancing these parameters is crucial in minimizing defects and ensuring the integrity of the weld joint [36-38].

The diversity of joints created is as varied as the materials being fused together. Different FSW joints demand tailored tool designs to address the unique challenges posed by specific applications. The lap joint, for instance, is a common configuration where two overlapping pieces are welded together. In lap joint applications, the FSW tool's geometry becomes crucial in ensuring uniform material flow and consolidation at the joint interface [34].

The butt joints, on the other hand, involve welding the ends of two abutting pieces. In such cases, the FSW tool's shoulder and pin design play a vital role in achieving proper penetration and consolidation, with considerations for minimizing root defects and achieving a smooth weld surface [37]. For T-joints, where one workpiece intersects another, the FSW tool design needs to accommodate the varying thicknesses and thermal gradients. Adaptive features in the tool, such as adjustable pin lengths or shoulder configurations, are instrumental in achieving consistent weld quality across the joint. Complex joints, like those encountered in welding dissimilar materials, demand specialized FSW tool designs. The challenge here lies in accommodating the differing thermal expansion rates and material properties. Adaptive tool features, combined with precise control over welding parameters, enable the creation of defect-free joints in dissimilar material applications. The significance of FSW tool design in these diverse joint configurations cannot be overstated. It goes beyond merely facilitating the welding process; it directly influences the structural integrity, mechanical properties, and performance of the final joint. For instance, in aerospace applications where joints undergo stringent quality requirements, the FSW tool's ability to minimize defects and maintain consistent material flow becomes paramount [36-38].

The comprehensive design of FSW accounts for a myriad of factors, including material properties, cooling mechanisms, adaptive features, precision systems, and defect mitigation strategies. This holistic approach ensures the success of the FSW process across diverse materials and applications, highlighting the adaptability and efficiency of this groundbreaking welding technique. Moreover, the significance of FSW tool design transcends a one-size-fits-all paradigm, dynamically adjusting to the nuanced demands of various joint configurations. The capacity to tailor the tool to specific applications underscores the versatility and efficacy of FSW across a spectrum of materials and welding scenarios, solidifying its position as a transformative technology in the realm of modern fabrication.

3. SIMILAR AND DISSIMILAR METAL JOINING

In spite of its advantages, the FSW method still results in changes to the microstructure, leading to diminished mechanical properties and reduced resistance to corrosion in the welded joints [39]. In particular, when it comes to age hardenable Al7xxx alloys, where the base materials are typically hardened to T7x conditions prior to undergo the FSW process, these joints tend to display increased vulnerability to localized corrosion (LC) compared to the base materials [40-41]. The post treatment on the friction stir welded plates is observed to be a fruitful approach in mitigation of the corrosion vulnerability. The solid-solution treated Al7050-T76 plates were joined with FSW and post treated with T76 heat treatment again and the corrosion behavior was evaluated. For the Al7050-T76 base metal alloy joints without any post heat treatment, it was observed that $MgZn_2$ η and ή phases have fragmented and dissolved in nugget zone and precipitates coarsening took place in heat affected zone, which led to larger changes with nugget, heat affected and thermo mechanically affected zones. Now, after post treatment with T76, it was observed that the phenomenon of $MgZn_2$ η and ή phases coarsening and solute segregation were suppressed effectively. Thus the welded all regions microstructure differences were not too intense, which led to improved corrosion resistance [42].

This research is also focused on marine offshore application and copper nickel alloy welding with steel. In conventional welding, copper and copper alloys are difficult to weld because of their high thermal conductivity and high melting point. This may lead to loss of its structural properties and strength. To overcome these problems, FSW is a better alternative for fusion/conventional welding.

There are efforts by different methods in order to improve corrosion and anti biofouling behavior [43-45]. The present practices followed for marine environment are: The cooper-nickel alloys have excellent corrosion resistance and inherent resistance to biofouling. Alloys with high nickel content are used where there is greater resistance to flow conditions, wear, and galling. There are no efforts reported on solid-state joining of these copper nickel alloys using friction stir welding.

This fouling happens mostly in warm conditions and to structures that have low velocity (<1m/s) or are stagnated. Removal of this fouling by mechanical means is required in such cases but to prevent it in the first place, various coatings of chlorine is used. Thus the joining of copper nickel alloy with steel alloys needs to be carried out using FSW in order to investigate the behavior of weld bead in marine environments. In some cases, the ocean bio contaminants attach themselves to the metal alloys more easily and are hard to remove. Metals such as steel, titanium, and aluminium are vulnerable to foul. But biofouling has been found to have more adherence to copper-based alloys. Copper nickel alloys have very good resistance with respect to biofouling. The 90-10 copper nickel alloy used in many marine structural applications that are resistant to bio fouling [43-45].

The dissimilar metal joining through FSW has been classified depending on its intensity of the interfacial reactions during the friction stir welding. The Al-Mg dissimilar welding is considered to be severe interfacial reaction welding. The Mg-Steel dissimilar joining is a non-interfacial reaction welding system. Al-Ti and Al-Steel dissimilar joining comes under the category of the mild interfacial reaction welding system [46]. The vital significance of joining dissimilar materials lies in its ability to enhance technical performance and contribute to the commercial success of a design. Over the past few decades, researchers have extensively studied the process of joining dissimilar materials. Many applications necessitate this type of joining to harness the desirable qualities of each material involved [47].

Numerous studies have demonstrated the effective application of solid-state joining processes in dissimilar material joining. Among these processes, FSW stands out as a prominent method for achieving successful dissimilar joining. The heat generated during the process reaches approximately 60–80% of the base metal's melting point. The solid-state welding characteristic of this method reduces the occurrence of solidification cracking and weld joint shrinkage. The FSW process parameters, including tool rotational speed and plunge depth, have a vital role in ensuring the appropriate heat generation. Suboptimal heat input, either too low or too high, can adversely affect the weld joint quality. Thus, optimizing the FSW process parameters becomes crucial to achieve well-formed and robust weld joints. Due to disparities in their physical and chemical properties, conventional fusion welding methods face significant challenges in welding brass and aluminum together, making it practically infeasible to achieve a strong weld between these two materials.

4. NUMERICAL TECHNIQUES IN FSW

FSW procedures need extensive study, optimization, and design using numerical methods. These methods include the use of mathematical simulations and models to comprehend the intricate thermal, mechanical, and metallurgical processes that take place during the actual process. Researchers and engineers can forecast the final weld quality, optimize the welding conditions, and even assist in the design of FSW tools and equipment by modeling the process. Finite element analysis (FEA), which discretizes the FSW/FSP domain into tiny elements and solves the governing equations for heat transport, material flow, and mechanical deformation, is one of the frequently used numerical techniques [48]. The temperature distribution, plastic flow, and residual stresses during FSW may all be precisely described using FEA. Predicting defects, such as voids or fractures, and optimizing process variables to enhance weld quality both benefit from this knowledge. Figure 8.3 illustrates the potential way of finite element modeling in FSW process.

Another numerical method used in FSW to examine material flow and heat transmission inside the weld zone is computational fluid dynamics (CFD). Understanding the patterns of material flow, the emergence of vortices, and the distribution of temperature gradients may be aided by CFD models.

Using this information, process parameters may be optimized to reduce or completely remove flaws like tunnel defects or flash. A validated couple-thermo flow

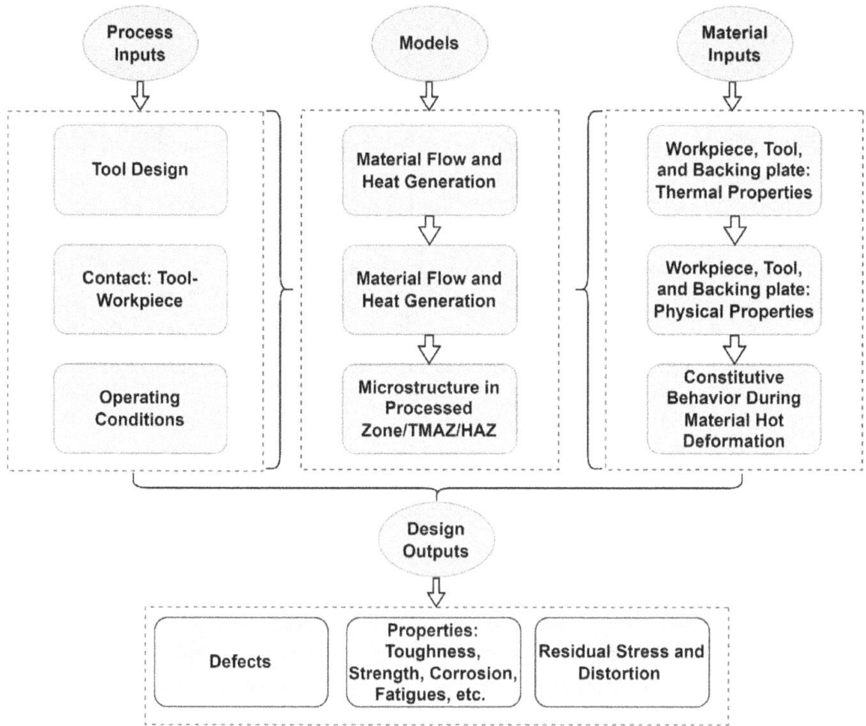

FIGURE 8.3 Finite element modeling of FSW

model was developed to simulate flash formation during FSW. The two-phase flow model predicted reduced pressure values on the tool surface, while the volume fraction contour showed changes in the metal phase with increased welding time. This study provides insights into flash formation phenomena and offers potential techniques to improve tool life and prevent tool failure in FSW [49].

In addition, the metallurgical changes that take place during the FSW process are simulated using phase transformation models and material constitutive equations. These models are able to forecast changes in the weld zone's microstructure, grain size, and mechanical characteristics. Understanding the material behavior and functionality of FSW joints requires knowledge of these information. Smoothed Particle Hydrodynamics (SPH) recently has taken interest by several researchers [38-39]. Additionally, experimental data and numerical methodologies may be used to test and enhance the models' correctness. The numerical models may be calibrated to assure accuracy and the dependability of the results using data from temperature sensors, force sensors, and non-destructive testing techniques.

In addition to experimental approaches, many efforts are taken to simulate the plasticized material flow during FSW using different finite element models, fundamental geometrical models, the boundary layer flow phenomena [50-52]. These initiatives were undertaken in an attempt to understand the fundamentals of the material flow behavior during friction welding. Plastic deformation causes heat production, and friction, which controls flow stress during FSW/ FSP, is essential for material flow.

Due to the inherent computing challenges, it is exceedingly challenging to address the complex, highly-coupled, nonlinear processes by computer-based numerical modeling as well as analysis of friction stir welding and processing [53]. Numerous sensitive components also increased complexity of the numerical analysis of friction stir welding using the finite elements method. The balance of energy and momentum is controlled by the same analogous equations, which describe thermo-mechanical connections of the problem. The nonlinear behavior of the control equations has a significant impact on the complex nature of models. Finding a robust numerical approach is thus essential to solve such a highly-coupled nonlinear finite element model. The physical FSW process may be mathematically modeled using following approaches, that is, the localized technique or the global method [54]. The heat affected zone of the analyzed workpiece is considered vital in the context of localized-level study, wherein the analysis aims to ascertain the frictional visco-plastic dissipation heat generated at the contact phases.

Among the most important friction stir physical process behaviors at this level are the interactions between process parameters and the contact interface mechanisms as they are directly affected by frictional coefficient, axial load, dimensions of FSW setup, material flow, weld material size, and the resultant microstructure. A global simulation is being used to analyze the whole component that will be processed [55-56]. The boundary condition that represents the instantaneous heat affected zone condition is controlled by the dynamic heat generation.

5. POSSIBLE FUTURE WORK AVENUES

FSW offers numerous advantages for joining materials and has been primarily used for aluminum alloys. However, future work avenues include expanding its application to new materials like high-strength steels, titanium alloys, and magnesium alloys, as well as optimizing FSW parameters to enhance weld quality and mechanical properties. Hybrid joining technologies, combining FSW with other techniques, can be explored for improved joint strength and productivity. Advanced tooling and equipment, including intelligent tooling and better process control, can be developed. Joining dissimilar materials, assessing structural integrity and fatigue performance, exploring additive manufacturing with FSW, and establishing design guidelines are important areas for further research and development. These prospective directions for future study show the potential for improvements in FSW and design, providing chances to streamline the procedure, raise joint quality, increase material capabilities, and investigate cutting-edge uses across sectors. These attempts will contribute to the continued advancement and adoption of FSW technology.

6. SUMMARY

This book chapter provides an in-depth analysis of FSW by covering a variety of subjects, such as joining metals that are similar and different, numerical methods to FSW, and likely future research areas. The emphasis of the chapter is on the advantages of FSW over conventional fusion welding techniques, including reduced distortions, prevention of solidification cracking, and improved weld quality. It delves into the challenges and considerations of mixing similar metal alloys with FSW and addresses process variables, tool design, and optimization techniques. Additionally, it looks at the challenges of joining metals with various chemical compositions, including intermetallic formation, metallurgical compatibility, and workable solutions for creating strong welds. The chapter also examines the potential for numerical techniques like FEA and CFD to mimic and enhance the FSW process. It finishes by detailing prospective research directions, including hybrid FSW processes, advancements in tool materials, techniques for in-situ monitoring, and improved numerical modeling and simulation.

Overall, for academicians, engineers, and practitioners looking to get a thorough grasp of FSW, this chapter is an invaluable resource. It examines the use of numerical tools for process optimization and identifies possible directions for future study. It offers insights into the difficulties and possibilities in comparable and dissimilar metal joining. The chapter advances FSW technology by addressing these issues and makes it easier for different sectors to use it.

REFERENCES

[1] W.M. Thomas, E.D. Nicholas, J.C. Needham, M.G. Murch, P. Temple-Smith, C.J. Dawes, Friction-Stir Butt Welding, GB Patent No. 9125978.8, International patent application No. PCT/ GB92/02203, 1991.

[2] R. Nandan, T. DebRoy, H.K.D.H. Bhadeshia, Recent advances in friction-stir welding - Process, weldment structure and properties, *Prog. Mater. Sci.* 53 (2008) 980–1023. https://doi.org/10.1016/j.pmatsci.2008.05.001

[3] R.S. Mishra, Z.Y. Ma, Friction stir welding and processing, *Mater. Sci. Eng. R Reports.* 50 (2005) 1–78. https://doi.org/10.1016/j.mser.2005.07.001

[4] P.L. Threadgilll, A.J. Leonard, H.R. Shercliff, P.J. Withers, Friction stir welding of aluminium alloys, *Int. Mater. Rev.* 54 (2009) 49–93. https://doi.org/10.1179/174328009X411136

[5] Z.Y. Ma, A.H. Feng, D.L. Chen, J. Shen, Recent advances in friction stir welding/processing of aluminum alloys: Microstructural evolution and mechanical properties, *Crit. Rev. Solid State Mater. Sci.* 43 (2018) 269–333. https://doi.org/10.1080/10408436.2017.1358145

[6] D.K. Rajak, D.D. Pagar, P.L. Menezes, A. Eyvazian, Friction-based welding processes: Friction welding and friction stir welding, *J. Adhes. Sci. Technol.* 34 (2020) 2613–2637. https://doi.org/10.1080/01694243.2020.1780716

[7] R.S. Mishra, Z.Y. Ma, Friction stir welding and processing, *Mater. Sci. Eng. R Reports.* 50 (2005) 1–78. https://doi.org/10.1016/j.mser.2005.07.001

[8] B. Abnar, S. Gashtiazar, M. Javidani, Friction stir welding of non-heat treatable Al alloys: Challenges and improvements opportunities, *Crystals.* 13 (2023). https://doi.org/10.3390/cryst13040576

[9] A. Heidarzadeh, S. Mironov, R. Kaibyshev, G. Çam, A. Simar, A. Gerlich, F. Khodabakhshi, A. Mostafaei, D.P. Field, J.D. Robson, A. Deschamps, P.J. Withers, Friction stir welding/processing of metals and alloys: A comprehensive review on microstructural evolution, *Prog. Mater. Sci.* 117 (2020) 100752. https://doi.org/10.1016/j.pmatsci.2020.100752

[10] T. Dursun, C. Soutis, Recent developments in advanced aircraft aluminium alloys, *Mater. Des.* 56 (2014) 862–871. https://doi.org/10.1016/j.matdes.2013.12.002

[11] C.A. De Saracibar, Challenges to be tackled in the computational modeling and numerical simulation of FSW processes, *Metals (Basel).* 9 (2019) 573. https://doi.org/10.3390/met9050573

[12] C.B. Smith, G.B. Bendzsak, T.H. North, J.F. Hinrichs, J.S. Noruk, R.J. Heideman, *Heat and Material Flow Modeling of Friction Stir Welding Process*, 2000.

[13] L. Long, G. Chen, S. Zhang, T. Liu, Q. Shi, Finite-element analysis of the tool tilt angle effect on the formation of friction stir welds, *J. Manuf. Process.* 30 (2017) 562–569. https://doi.org/10.1016/j.jmapro.2017.10.023

[14] D.G. Mohan, C.S. Wu, A review on friction stir welding of steels, *Chinese J. Mech. Eng. (English Ed.* 34 (2021). https://doi.org/10.1186/s10033-021-00655-3

[15] S. Guerdoux, L. Fourment, A 3D numerical simulation of different phases of friction stir welding, *Model. Simul. Mater. Sci. Eng.* 17 (2009). https://doi.org/10.1088/0965-0393/17/7/075001

[16] X. He, F. Gu, A. Ball, A review of numerical analysis of friction stir welding, *Prog. Mater. Sci.* 65 (2014) 1–66. https://doi.org/10.1016/j.pmatsci.2014.03.003

[17] J.Q. Su, T.W. Nelson, C.J. Sterling, Microstructure evolution during FSW/FSP of high strength aluminum alloys, *Mater. Sci. Eng. A.* 405 (2005) 277–286. https://doi.org/10.1016/j.msea.2005.06.009

[18] R.V. Marode, S.R. Pedapati, T.A. Lemma, V. Somi, R. Janga, Thermo-Mechanical modelling of friction stir processing of AZ91 Alloy: Using smoothed-particle hydrodynamics, *Lubricants.* 10 (2022) 1–20. https://doi.org/10.3390/lubricants10120355

[19] R.S. Mishra, Z.Y. Ma, Friction stir welding and processing, *Mater. Sci. Eng. Rep.* (2005). https://doi.org/10.1016/j.mser.2005.07.001

[20] X. Wang, Y. Gao, X. Liu, M. McDonnell, Z. Feng, Tool-workpiece stick-slip conditions and their effects on torque and heat generation rate in the friction stir welding, *Acta Mater.* 213 (2021) 1359-6454. https://doi.org/10.1016/j.actamat.2021.116969

[21] L. Shi, J. Chen, C. Yang, G. Chen, C. Wu, Thermal-fluid-structure coupling analysis of void defect in friction stir welding, *Int.J. Mech. Sci.* 241 (2023) 107969. https://doi.org/10.1016/j.ijmecsci.2022.107969

[22] S. Li, H. Dong, L. Shi, P. Li, F. Ye, Corrosion behavior and mechanical properties of Al-Zn-Mg aluminum alloy weld, *Corros. Sci.* 123 (2017) 243-255.https://doi.org/10.1016/j.corsci.2017.05.007

[23] P.S. Davies, B.P. Wynne, W.M. Rainforth, M.J. Thomas, P.L. Threadgill, Development of microstructure and crystallographic texture during stationary shoulder friction stir welding of Ti-6Al-4V, *Metall. Mater. Trans. A* 42 (2011) 2278–89. https://doi.org/10.1007/s11661-011-0606-2.156

[24] H. Zhang, M. Wang, W. Zhou, X. Zhang, Z. Zhu, T. Yu, G. Yang, Microstructure–property characteristics of a novel non-weld-thinning friction stir welding process of aluminum alloys, *Mater. Des.* 86 (2015) 379–387. https://doi.org/10.1016/j.matdes.2015.06.106

[25] H. Wu, Y.C. Chen, D. Strong, P. Prangnell, Stationary shoulder FSW for joining high strength aluminum alloys, *J. Mater. Process Technol.* 221 (2015) 187–196. https://doi.org/10.1016/j.jmatprotec.2015.02.015

[26] P.L. Threadgill, A.J. Leonard, H.R. Shercliff, P.J. Withers, Friction stir welding of aluminium alloys, *International Materials Reviews* 54 (2009) 49–93.

[27] W.F. Xu, H.J. Lu, X.H. Li, M. Wang, J. Ma, Y.X. Luo, Microstructure evolution and stress corrosion cracking sensitivity of friction stir welded high strength AA7085 joint, *Mater. Des.* 212 (2021). https://doi.org/10.1016/j.matdes.2021.110297

[28] Y. Wang, Y. Chen, L. Zhou, Y. Shao, L. Liu, J. Jiang, Improving the corrosion resistance of friction stir welding joint of 7050 Al alloy via optimizing the process route, *J. Mater. Res. Techn.* 24 (2023) 8098–8108. https://doi.org/10.1016/j.jmrt.2023.05.053

[29] W.Y. Li, T. Fu, L. Hutsch, J. Hilgert, F.F. Wang, J.F. Dos Santos, Effects of tool rotational and welding speed on microstructure and mechanical properties of bobbin-tool friction-stir welded Mg AZ31, *Mater. Des.* 64 (2014) 714–20. https://doi.org/10.1016/j.matdes.2014.07.023

[30] C. Schilling, J. Dos Santos, Method and device for linking at least two adjoining work pieces by friction welding, 2001. Patent No. US6722556B2.

[31] E. Hoyos, M.C. Serna, Basic tool design guidelines for friction stir welding of aluminum alloys, *Metals* 11 (2021) 2042. https://doi.org/10.3390/met11122042

[32] R. Rai, A. De, H.K.D.H. Bhadeshia, T. DebRoy, Review: Friction stir welding tools, *Sci. Technol. Weld. Join.* 16 (2011). https://doi.org/10.1179/1362171811Y.0000000023

[33] W.H. Jiang, R. Kovacevic, Feasibility study of friction stir welding of 6061-T6 aluminium alloy with AISI 1018 steel: Proceedings of the Institution of Mechanical Engineers, *Part B: Journal of Engineering Manufacture.* (2005).

[34] A. Elrefaey, M. Gouda, M. Takahashi, Characterization of Aluminum/Steel lap joint by friction stir welding, *J. Mater. Eng. Perform.* 14 (2005) 10–17. https://doi.org/10.1361/10599490522310

[35] H. Uzun, C. Dalle Donne, A. Argagnotto, Friction stir welding of dissimilar Al 6013-T4 To X5CrNi18-10 stainless steel. *Mater. Des.* 26 (2005) 41–46. https://doi.org/10.1016/j.matdes.2004.04.002

[36] Z.W. Chen, S. Yazdanian, Friction stir lap welding: Material flow, joint structure and strength. *J. Achiev. Mater. Manuf. Eng.* 9 (2012) 55.

[37] C. Bitondo, U. Prisco, A. Squilace, Friction-stir welding of AA 2198 butt joints: mechanical characterization of the process and of the welds through DOE analysis. *Int. J. Adv. Manuf. Technol.* 53 (2011) 505–516. https://doi.org/10.1007/s00170-010-2879-9

[38] R.V. Marode, M. Awang, V.S.R. Janga, Computational modelling and comparative analysis of friction stir welding and stationary shoulder friction stir welding on AA6061. *Crystals* 13 (2023) 1317. https://doi.org/10.3390/CRYST13091317

[39] Z. Shen, Y. Ding, O. Gopkalo, B. Diak, A.P. Gerlich, Effects of tool design on the microstructure and mechanical properties of refill friction stir spot welding of dissimilar Al alloys, *J. Mater.Process Technol.* 252 (2018) 751–9. https://doi.org/10.1016/j.jmatprotec.2017.10.034

[40] Z. Xu, Z. Li, S. Ji, L. Zhang, Refill friction stir spot welding of 5083-O aluminum alloy, *J. Mater. Sci. Technol.* 34 (2018) 878–85. https://doi.org/10.1016/j.jmst.2017.02.011

[41] R.V. Marode, S.R. Pedapati, T.A. Lemma, M. Awang, A review on numerical modelling techniques in friction stir processing: current and future perspective, *Arch. Civ. Mech. Eng.* 23 (2023). https://doi.org/10.1007/s43452-023-00688-6

[42] X. He, F. Gu, A. Ball, A review of numerical analysis of friction stir welding, *Prog. Mater. Sci.* 65 (2014) 1–66. https://doi.org/10.1016/j.pmatsci.2014.03.003

[43] R. Rajesh, M. V. Kulkarni, B. R. Vergis, P. Sampathkumaran, S. Seetharamu, Copper nickel alloys made using direct metal laser sintering method for assessing corrosion resistance properties, *Mater. Today: Proc.* 49 (2022) 703–713. https://doi.org/10.1016/j.matpr.2021.05.178

[44] S.P. Vinodhini, J.R. Xavier, Evaluation of newly synthesized multifunctional nanocomposite coated cupronickel alloy in marine environment, *Mater. Chem. Phys.* 268 (2021). https://doi.org/10.1016/j.matchemphys.2021.124721

[45] G. Jena, B. Anandkumar, S.C. Vanithakumari, R.P. George, J. Philip, G. Amarendra, Graphene oxide-chitosan-silver composite coating on Cu-Ni alloy with enhanced anticorrosive and antibacterial properties suitable for marine applications, *Prog. Org. Coat.* 139 (2020), https://doi.org/10.1016/j.porgcoat.2019.105444

[46] S. Sen, J. Murugesan, Experimental and numerical analysis of friction stir welding: A review, *Eng. Res. Express*. 4 (2022) 032004. https://doi.org/10.1088/2631-8695/ac7f1e

[47] W.M. Thomas, P.L. Threadgill, E.D. Nicholas, Feasibility of friction stir welding steel, *Sci. Technol. Weld. Join.* 4 (1999) 365–372. https://doi.org/10.1179/136217199101538012

[48] A.F. Hasan, CFD modelling of friction stir welding (FSW) process of AZ31 magnesium alloy using volume of fluid method, *J. Mater. Res. Technol.* 8 (2019) 1819–1827. https://doi.org/10.1016/j.jmrt.2018.11.016

[49] Y. Yin, X. Yang, L. Cui, F. Wang, S. Li, Material flow influence on the weld formation and mechanical performance in underwater friction taper plug welds for pipeline steel, *Mater*. Des.88 (2015) 990–8. https://doi.org/10.1016/j.matdes.2015.09.123

[50] X. Meng, Y. Huang, J. Cao, J. Shen, J.F. Dos Santos, Recent progress on control strategies for inherent issues in friction stir welding, Prog. *Mater. Sci.* 115 (2021), 100706, ISSN 0079 6425, https://doi.org/10.1016/j.pmatsci.2020.100706

[51] R.V. Marode, M. Awang, T.A. Lemma, S.R. Pedapati, A. Hassan, V.S. Reddy Janga, Numerical analysis of temperature and material flow predictions with defects in the friction stir processing of AZ91 alloy: An advanced meshfree SPH technique, *Eng. Anal. Bound. Elem.* 161 (2024) 48–69. https://doi.org/10.1016/j.enganabound.2024.01.016

[52] R.V. Marode, S.R. Pedapati, T.A. Lemma, S.S.K. Bellala, Investigation on AA7075 and AA7075/SiC surface composites using friction stir processing: An experimental and numerical approach, in: Mater. Today Proc., Elsevier Ltd, 2023. https://doi.org/10.1016/j.matpr.2023.04.428

[53] P. Dong, F. Lu, J.K. Hong, Z. Cao, Coupled thermomechanical analysis of friction stir welding process using simplified models, *Sci. Technol. Weld. Join.* 6 (2001) 281–287. https://doi.org/10.1179/136217101101538884

[54] H. Mehdi, R.S. Mishra, Analysis of material flow and heat transfer in reverse dual rotation friction stir welding: A review, *Int. J. Steel Struct.* 19 (2019) 422–434. https://doi.org/10.1007/s13296-018-0131-x

[55] N. Dialami, M. Chiumenti, M. Cervera, C. Agelet de Saracibar, Challenges in thermo-mechanical analysis of friction stir welding processes, *Arch. Comput. Methods Eng.* 24 (2017) 189–225. https://doi.org/10.1007/s11831-015-9163-y

[56] C.A. De Saracibar, Challenges to be tackled in the computational modeling and numerical simulation of FSW processes, *Metals (Basel)*. 9 (2019) 573. https://doi.org/10.3390/met9050573

ABBREVIATIONS

FSW	Friction stir welding
FSSW	Friction stir spot welding
NZ	Nugget zone
HAZ	Heat affected zone
TMAZ	Thermo mechanically affected zone

9 Laser Beam Welding: Opportunity and Challenges

Atul Kumar, Anoj Giri, Ashish Kumar Saxena, and Rishav Kumar

1. INTRODUCTION

In recent scenarios, lasers are used in various engineering domains such as material joining machining, laser engraving, additive manufacturing, and so on. In the manufacture of high-performance materials, alloy steels, composite materials, super strong stainless steels, titanium alloys, and magnesium alloys, consideration shall be given to the environmental compatibility of high wear resistance coating and other products, as well as their interrelation with welding techniques. Laser welding has been considered an ecological activity, given that its emissions are very low as compared to arc welding. Laser welding is the prime method of welding used in industry with conventional laser sources. It allows better energy utilization and larger depth-to-width ratio welds are attainable with deep penetration welding [1-2].

The unique characteristics of laser beams, such as high monochromaticity and extremely intense and directional beams of electromagnetic radiation, enable them to be used in welding, metallurgy, and other forms of material processes. Primarily there are three classes of lasers, that is, solid, liquid, and gas laser. All classes of lasers operate in one of the two quantum states: continuous wave and pulse wave form. Some other types of lasers are also used for commercial applications for many devices, those are known as chemical lasers, metal-vapor lasers and semiconductor lasers. The fiber lasers can be applied to a wide range of applications, for example from joining extremely small objects used in the medical engineering and electronics industries to welding thick components in the manufacturing of vehicles and aircraft.

The laser beam facilitates a concentrated light beam which works as a heat source, allowing high depth-to-width ratio during welding. In welding, a deep weld pool is obtained, and the applied heat input to the joint is very low compared to conventional welding process. The joints with minimum distortion can also be produced by laser beam welding. The aesthetics, elimination of the machining process, and capacity to weld inaccessible locations have piqued the interest of industries.

DOI: 10.1201/9781003435884-9

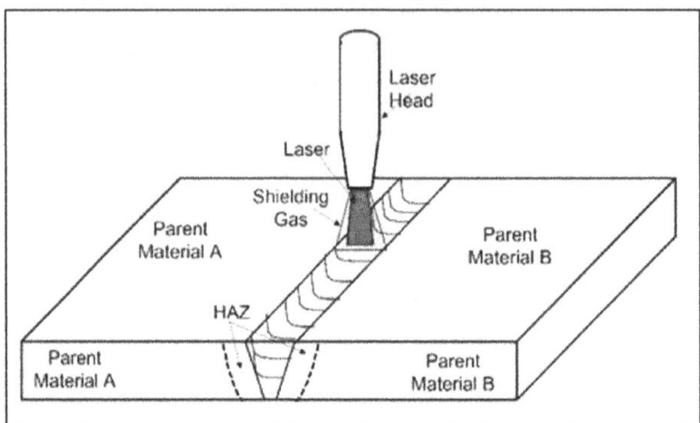

FIGURE 9.1 Schematic of laser beam welding process

2. LASER BEAM WELDING FOR MATERIAL JOINING

Laser beam welding (LBW) addresses the issues such as wider HAZ, excessive distortion, higher heat input, and undesirable metallurgical changes encountered in the traditional welding process, and adds advantages such as lower heat input, smaller heat-affected zone, flux elimination, reduced weld distortion, faster welding speed, and minimum degradation of heat sensitive material, and so on [1].

For the joining process, LBW is a high-energy density fusion welding process, in which the heat required to melt the edge is obtained by irradiating a high-energy laser beam [3]. In LBW, the absorption of intensified radiation has created the molten weld pool region. The constant absorption of the laser radiation, as well as a consistent distribution of heat in the workpieces, plays an essential part in balancing thermal input and output [3]. The high aspect ratio weld can be made by LBW. In the laser welding process, the weld penetration depth and width can be adjusted by controlling the process parameters such as: laser power, focal position, welding speed, energy mode (pulsed or continuous mode) and laser spot size. If the melting area is too large, or too small, or material is overly evaporated, it may result in a failure of joints. The weld quality shall be analyzed based on the evaporation of the alloying elements and the thermal gradient of the phenomena leading to the formation of cracks, as described in the following paragraphs. The schematic of laser welding is shown in Figure 9.1.

Laser welding can be performed in mainly two modes: conduction and keyhole techniques, and each produces a different weld bead profile.

2.1. Conduction mode

Laser welding at lower energy density results in a shallow weld profile. The maximum weld depth in conduction mode welding, which normally ranges from only a few tenths of a millimeter to 1 millimeter, is constrained by the heat conductivity of the materials. Energy is linked to the workpiece only through heat conduction. Conduction welding is ideally suited for joining thin-wall materials and spot welding since the breadth of a conduction weld is always greater than the depth of the joint

[4]. The electronics packaging and corner welds are created via conduction mode welding. Battery sealing is another typical conduction mode weld application. Heat conduction welding is also suitable for volatile alloys such as magnesium and zinc, where keyhole welding is not preferred [5].

2.2. Keyhole Mode

The term "keyhole" refers to a physical hole created in the substance by its vaporization, which enables the energy beam to travel further. Increasing the energy density, the shallow keyhole mode is changed into deeper keyhole mode. The deeper narrow welds have an aspect ratio larger than 1:5 can be achieved by this technique. The keyhole technique is useful in case of thick plates welding, where the deeper weld penetration is primary objective. Keyhole mode welds are recommended for structural and deep penetration welds. Keyhole mode welding is used in industries, for example, turbine blades, space-bound valves, tank bearings. The schematic diagram of conduction and keyhole laser welding modes is shown in Figure 9.2.

In contrast to arc welding, which produces penetration by increasing the heat input, LBW is a high-energy density approach that does not require thermal conduction to

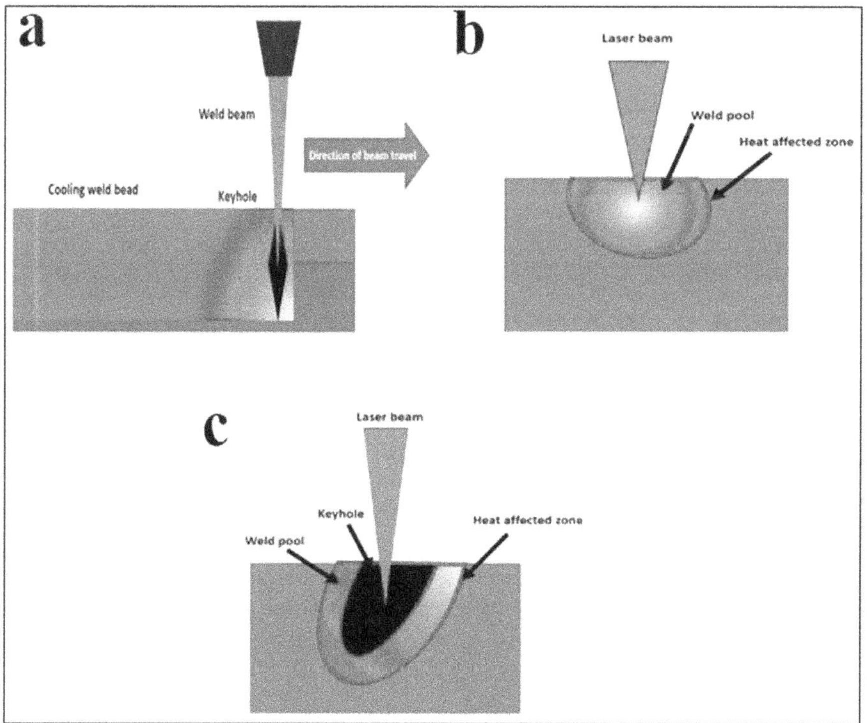

FIGURE 9.2 Laser welding mechanisms: (a) laser welding, (b) conduction mode & (c) keyhole mode

TABLE 9.1
Conduction and keyhole laser welding

	Conduction laser welding	Keyhole laser welding
Heat efficiency	Approx. 35%	Approx. 95%
Tolerance	Accuracy required on both surface	Accuracy required only on one surface
Efficiency	Low	High
Welding speed	Slow	Fast
Energy required	High	Low
Distortion	Slight distortion	Minimal distortion
Seam size	Large	Small
Seam smoothness	Smooth	Rough

create a deep penetration. When using the LBW, a deep and narrow weld pool is created due to the keyhole phenomenon. The comparison between the conduction laser welding and keyhole laser welding is shown in Table 9.1.

3. ROLE OF LASER BEAM WELDING PROCESS PARAMETERS

Laser welding is mainly classified into two modes, that is, continuous mode and pulse mode. In continuous mode, the laser beam is radiated continuously while in pulse mode, the laser beam is periodic irradiated. In continuous wave laser welding, the welding parameters are laser power, welding speed, beam diameter, and the defocus amount. However, the parameters like pulse shape, pulse width, pulse duration, and frequency are also associated with the pulse laser welding. The heat input in pulse laser welding can be controlled (or minimized as compared to continuous laser welding) by adjusting the parameters simultaneously. The heat input increases either by increasing the laser power or decreasing the welding speed, which results in the increase of depth of penetration and weld width. Generally, with constant power, the weld pool is smaller, the heat is more concentrated, and the beam diameter decreases [3].

Typically, laser welding employs a square pulse form (as shown in Figure 9.3a). Spike pulses are light reflective, utilized for materials like copper and aluminium. The annealing pulse is deployed to reduce the radiant heat cycle during welding for crack-sensitive materials. To avoid puncturing the welding region, pre-pulses with a lower peak power are descended on thin foils before the main pulse.

Pulse width controls the heat input to the workpiece. Increasing the pulse width expands the weld width and HAZ width due to the increased heat transfer time, while the peak power controls the depth of penetration. The pulse width typically regulates the heat input to the workpiece, while the power density regulates the penetration of the weld.

The selection of laser source is dependent upon the material to be welded. Fiber lasers are versatile and inexpensive, which are suitable for achieving high-quality spot welds. The Nd: YAG pulsed laser produces the discrete pulse of control energy

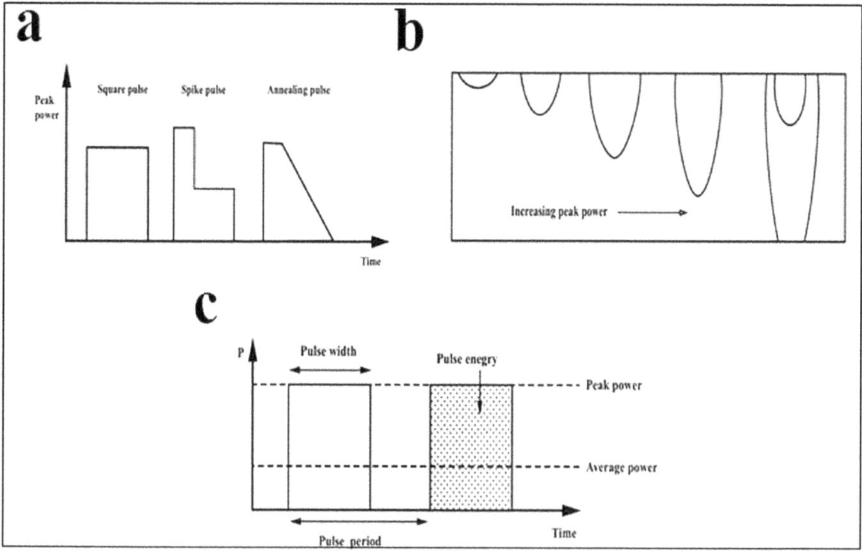

FIGURE 9.3 Laser welding parameters: (a) schematic of various pulse forms, (b) effect of peak power of depth of penetration and (c) pulse width and power peak

that is capable of being applied for a perfect weld. Moreover the Nd:YAG lasers outperform gas lasers in terms of efficiency, optical quality, and physical characteristics.

4. CHALLENGES IN LASER BEAM WELDING

4.1. Joining of Dissimilar Materials

The joining of dissimilar materials has become inevitable in the engineering sector. To do so, there is a constant effort to use dissimilar metal combinations notwithstanding several difficulties that have come about, including the problems with metallurgical incompatibility. In comparison with the arc welding and solid-state welding processes, the high-energy density welding processes like electron beam welding (EBW) and laser beam welding are still of great industrial interest due to the numerous advantageous such as versatility, high welding speed, controlled heat input, minimum distortion, and less damage to the parent metal [3]. In areas difficult to access using conventional welding techniques, laser beam welding can make successful connections between complicated joints. An accurate location for the joint is provided via optical seam tracking and laser triangulation during the welding.

4.2. Common Metallurgical Defects During Laser Welding

The penetration depth of laser beam limits the application of laser beam welding for most cases. Laser beam welding is excellent for thin sheets and components, but thicker section laser beam welding is not suitable.

The rapid heating and cooling technique used in laser welding is the most important characteristic. However, the faster heating and cooling rates lead to solidification cracking and porosity in the weld region. Intermetallic compound formation, crack and pore are some issues in laser welding which more frequently occurs due to rapid heating to high temperature. Different specimen locations result in various types of intermetallic compounds, they are of distinct sizes and quantities [6]. SEM examination may be used to determine the bonded zone's size and the intermetallic compound layer's thickness [7-9].

4.3. Development of residual stress

Nonuniform weld pool and base metal area expansion/ contraction during the welding procedure with different metals leads to deformation and eventually produces residual stresses. Excessive residual stress generation can result in distortions, early failures, fatigue cracking, and many other serious issues.

Researchers have demonstrated that a fiber laser beam with a center laser spot encircled by a doughnut can significantly increase the stability of a keyhole weld.

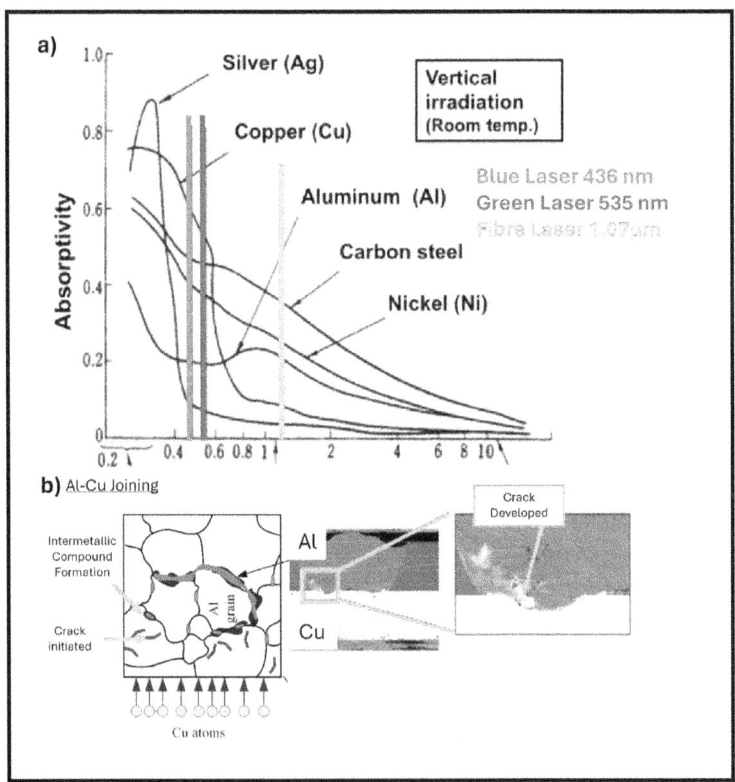

FIGURE 9.4 (a) Laser absorption efficiency of different metals, (b) crack formation in Al-Cu joints due different absorption efficiency for various wavelength laser beams [10]

This beam shape can improve the homogeneity of the fusion zone and joint strength by reducing spatter and porosity.

4.4. Laser sensitivity to materials

Those materials which are highly reflective in nature, for example, Cu, Ag, and Al, also have good thermal conductivity, creating problems in the laser welding process because a large amount of energy is reflected rather than absorbed by the materials for joining purposes. Moreover, whatever energy gets absorbed is quickly dispersed from the welding zone region [10]. Apart from that, aluminium and high carbon steel get distortion and cracking due to rapid heating and cooling cycles at the weldment

4.5. Safety concern

High-energy laser beams are generated in the whole system, which can be risky for human body and environment. Inappropriate handling of instruments may cause severe injuries, burning, or skin-related problems. On the other hand the entire process produces toxic fumes of vaporized particles that may cause problems for workers in the industry, thus strict safety protocols must be in place to deal with any such issues.

4.6. High-cost concern

The setup for the laser welding process requires a large initial investment to cope with advance optics and control requirement, as well as to fulfill the safety protocol for industry workers. On the other hand, the service and maintenance increase the cost of the laser processing of materials.

5. DISSIMILAR WELDING: EMERGING DEVELOPMENTS

5.1. Emerging advancements on the difficulties of welding steel with other alloys (Al, Mg, Ti)

Any welding process's efficiency is significantly influenced by the material's physical characteristics. Due to its great tensile strength and low cost, steel is an extensively utilized structural material. On the other hand, Al is available in adequate quantities and has a very low density, high corrosion resistance, and strong strength-to-ductility ratio. Steel and aluminum combined can offer reasonable cost, good formability, moderate strength, low density, and excellent corrosion resistance. The development of intermetallic compounds (IMCs) (mostly Al_xFe_y) during solidification is the fundamental technical problem of connecting steel-Al. In addition, there are notable variations between these two materials' thermal characteristics, such as melting point, thermal conductivity, coefficient of thermal expansion, and so on [11-13].

The welding of dissimilar materials is done using very specific requirements. The arc welding of dissimilar metallic materials is very challenging due to material incompatibility, filler material selection, and weld cracking during the welding. The laser welding of the dissimilar metals has attracted attention. The work related to the dissimilar metals joining by laser welding has been reported in the literature [14-21]. The dissimilar materials such as Zn coated steel/Al-alloy, SS 321/SS630, AISI 304/AISI 420 stainless steels [15], Superalloy K418/42CrMo steel [17], SS310/Inconel 657 [18], Carbon steel/Al-5754 alloy [19] and AZ31B/SS316 [20] have been successfully laser welded by various researchers.

In dissimilar welding, for example Steel /Al, Steel/Mg, and so on. the IMCs have a significant impact on the mechanical properties weld joint [21]. The interface needs to receive enough energy so that the surface is adequately wetted. The thickness and intermetallic compound layer are greatly influenced by power density; for a given laser spot energy, a greater power density causes a thicker IMC layer to develop [22]. The gases trapped in the solidifying weld pool are the primary cause of micro holes developing close to the fusion zone. Another factor may be the collapse of unstable laser welding keyholes [23].

6. APPLICATIONS OF LASER BEAM WELDING

Presently, the laser is used to join similar and dissimilar materials, even though the reflective metals are in similar and dissimilar conditions [24]. For various manufacturers, investment in laser beam welding is proven to be highly effective by reducing post-finishing processes and cycle time. Laser beam welding techniques are being used in modern industries [25].

In the automotive industry the laser beam welding technique is an effective and excellent choice to join various parts and accessories of an automobile together such as solenoids, suspension components, body panels, exhaust systems, air conditioning units, alternators, fuel injectors, and filters, and so on. Due to manufacturing efficiency, quality, and cost-effectiveness, laser welding has played a key role in the automobile industry during the past several years [26].

LBW is a widely used in the aerospace industry due to its precision in joining various aircraft parts, such as airframes, engine components, wings, fuselages, and so on. Fiber lasers are used to produce stronger and fatigue-resistant welds. In aerospace application, laser welding provides high precision, and low and localized heat input. Moreover, unlike conventional welding methods, a different material can be welded by laser beam welding process.

Nowadays, lasers are employed in several mobile welding applications since the process generates less overall heat, resulting in a smaller heat-affected zone (HAZ) and more localized heat. In the electronics industry, laser beam welding plays a vital role in joining components and parts of electronic devices together such as wires, connectors, sensors, PCBs, thermocouples, switches, relay terminals, lithium batteries, switches, and so on. The Nd:YAG or fiber laser is significantly used in the electronics sector. Precision welds for tiny, cutting-edge electrical components, for example control units and certain types of transistors, can be produced through LBW [26].

The LBW process does not create any contact or splatter of material during joining which lead to an extremely precise and clean weld. The medical sector frequently employs Nd:YAG and continuous wave lasers for micro welding, in which the welds are made to a depth of less than 1 mm. LBW is the ideal solution because it can produce incredibly small weld seams. Laser weld seams and spots may be produced even in complex locations, providing sterile surfaces without the need for post-processing.

LBW is a fast and efficient method that meets the high standards of quality and accuracy demanded by the medical industry. To join small and delicate parts of medical equipment and processes such as implants, prosthetics, endoscopic equipment, pacemakers hearing aids, and so on, LBW is a preferred choice in medical industries. The potential of LBW is successfully demonstrated in dental prosthesis [27], they have employed intraoral laser welding (with energy 9.9 mJ, spot size 0.6 mm and pulse frequency 1 Hz) for just 15 msec pulse duration for bar to dental implants of an elderly patient (Figure 9.5a intraoral laser welding of bar inserted in the abutments). This welding was completed in 47 seconds only, which has opened new opportunities of intraoral welding for dental application with limited anesthesia and better aesthetic and comfort level. It is also a cost-effective and quick technique. A wide variety of metals and alloys can be laser welded, without adding any filler materials, which avoids potential biocompatibility issues.

Additive manufacturing (AM) technology has transformed the world of manufacturing through its direct digital customization and geometrically intricate capabilities. LBW is one of the main technologies that has been used for advancement of AM's potentials. The laser welding system is used to selectively melt and fuse wire or powder-based materials to fabricate 3D objects. AM using laser power as a heat source can be classified as laser metal fusion and laser metal deposition (Figure 9.5 (b) the working procedure of laser metal fusion and (c)laser metal deposition processes).

Despite the number of benefits of AM technology, such as near-net-shape fabrication, shorter lead-time, AM technology has limitations in terms of component dimensions (due to the limited size of build chamber volume ~0.02 m^3), which limit its widespread adoption as an industrial processing route for engineering applications to a large extent [29-30]. Utilization of laser beam welding process could be the viable solution to join the small, printed part to fabricate the bulk structure with no dimensional limitations. Recently, some researchers have successfully joined additively manufactured materials, including 304 stainless steel, Al-Si alloy, and Ti-6Al-4V alloy through LBW processes [31-32]. Foued et al. [32] did the laser welding of AM fabricated 316L steel coupons and tested for high cycle fatigue. They concluded that laser welding does not deteriorate the fatigue properties of AM fabricated 316L specimens. It shows the potential of LBW for joining the additively produced components.

Even though the laser beam welding process is explored for various similar and dissimilar metallic systems, there is still scope for research to work for microstructure development correlation with the laser beam welding process. Nowadays the laser welding system has been improved thus able to control the welding parameters to greater extent so the microstructure can be tailored to restore or maintain the required mechanical properties within the components.

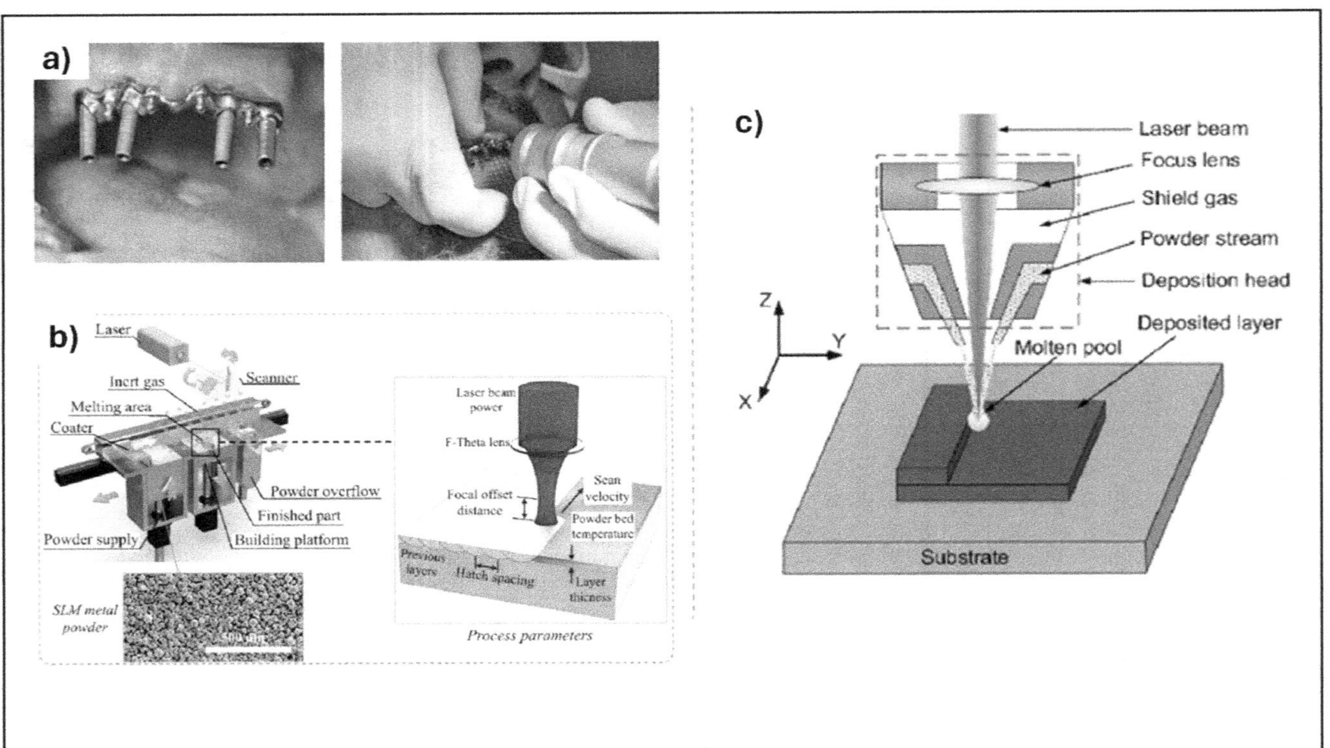

FIGURE 9.5 Laser beam welding applications for various purposes: (a) bar screwed in the abutments, intraoral laser welding process, (b) laser metal fusion, and (c) laser metal deposition processes [28]

7. CONCLUSIONS

This chapter highlights the laser beam welding process, the various factors effecting the quality of joining, the advancement, and challenges in laser welding. The LBW process is essential for the manufacturing industry, where the need for fast welding and defect-free welds are the primary objectives. Laser welding is being used over the arc welding processes due to less heat input and minimum HAZ weld characteristic. The control of the welding issues such as intermetallic formation, residual stresses, and distortion are of utmost importance during laser welding. Laser welding is advantageous for dissimilar material joining because of its precise and localized heat input, but suitable welding parameters are required for sound weld joints.

The laser welding of reactive materials and the highly reflective materials are challenging. Laser welding excels in automation and is suitable for a wide range of materials with a thickness range of 0.1 mm to 50 mm. Currently, laser power is also being used for additively manufactured automobile part production. Moreover, laser welding of dissimilar materials such as copper, aluminium, steel, nickel, titanium is still not explored, as they cannot be welded or are difficult to join by conventional joining techniques. Day by day the laser welding process is becoming versatile tool for various industries because of weld quality and operational speed.

REFERENCES

[1] Jayanthi, A., Kvenkataramanan, K. K., 2016. "Laser beams a novel tool for welding: A review," *IOSR Journal of Applied Physics (IOSR-JAP)*, 8(6) Ver. III.

[2] Walsh, C. A. "Laser Welding - Literature Review," England, University of Cambridge, 2002.

[3] Morteza Tayebi, Hedayat Mohammad Soltani, Ali Rajaee "Laser Welding," in *Engineering Principles - Welding and Residual Stresses*, Tehran, Iran, IntechOpen, 2022, pp. 3.

[4] "EB Industries," EB Industries, 2023. [Online Source] Available: https://ebindustries.com/conduction-mode-and-keyhole-mode-welding/

[5] "Laser welding techniques," KEYENCE, 2023. [Online Source]. Available: www.keyence.com/ss/products/measure/welding/laser/advanced.jsp

[6] Ding, H., Ma, J., Zhao, C., Zhao, D., 2021. "Effect of welding speed, pulse frequency, and pulse width on the weld shape and temperature distribution in dissimilar laser welding of stainless steel 308 and brass alloy," *Journal of. Laser Applications*, 2021, pp. 33.

[7] Meco, S., Pardal, G., Ganguly, S., Miranda, R. M., Quintino, L., Williams, S., 2012. "Overlap conduction laser welding of aluminium to steel," *Advance Manufacturing Technology*, pp. 67.

[8] Shi Yan, Zhang Hong, Takehiro Watanabe, Tang Jingguo, 2010. "CW/PW dual-beam YAG laser welding of steel/aluminum alloy sheets," *Optics and Lasers in Engineering*, 2010, pp. 732–736.

[9] Jin Yang, Yu-long Li, Hua Zhang, 2016. "Microstructure and mechanical properties of pulsed laser welded Al/steel dissimilar joint," *Trans. Nonferrous Metal Society of China, English* Ed. 2016, pp. 994–1002.

[10] Ma, B., Gao, X., Huang, Y., Gao, P. P., Zhang, Y., 2023. "A review of laser welding for aluminium and copper dissimilar metals," *Optics and Laser Technology*, 167, pp. 109721.

[11] Kotadia, H. R., Franciosa, P., Ceglarek, D., "Challenges and opportunities in remote laser welding of steel to aluminium," in MATEC Web of Conferences, 2019.

[12] Phanikumar, G., Chattopadhyay, K., Dutta, P., 2011. "Joining of dissimilar metals: Issues and modelling techniques," *Science and Technology of Welding and Joining*, pp. 313–317.

[13] Suryakanta Sahu, Suriya Kanta Pal, Mahadev Shome, Prakash Srirangam, 2021. "Welding of dissimilar metals—challenges and a way forward with friction stir welding" *Welding Technology*, pp. 167–192.

[14] Anawa, E. M., Olabi, A., 2008. "Optimization of tensile strength of ferritic/ austenitic laser welded components". *Optics and Lasers in Engineering*, 46(8), pp. 571–577.

[15] Roberto, B. J., Wagner, de R., Martin das, N., Ivan Alves de, A., Nilson Dias, V. J., 2007. "Pulsed Nd:YAG laser welding of AISI 304 to AISI 420 stainless steels" *Optics and Lasers in Engineering*, 45(9), pp. 960–966.

[16] Chengwu, Y., Binshi, X., Xiancheng, Z., Jian, H., Jun, F., et al. 2009. "Interface microstructure and mechanical properties of laser welding" *Optics and Lasers in Engineering*, 47, pp. 807–814.

[17] Xiu-Bo, L., Gang, Y., Ming, P., Ji-Wei, F., Heng-Hai, W., et al. 2007. "Dissimilar autogenous full penetration welding of superalloy K418 and 42CrMo steel by a high power CW Nd:YAG laser" *Applied Surface Science*, 253(17), pp. 7281–7289.

[18] Naffah, H, Shamanian, M., Ashrafizadeh, F., 2009. "Dissimilar welding of AISI 310 austenitic stainless steel to nickel-based alloy Inconel 657", *Journal of Materials Processing Technology*, 209(7), pp. 3628–3639.

[19] Torkamany, M. J., Tahamtan, S., Sabbaghzadeh, J., 2010. "Dissimilar welding of carbon steel to 5754 aluminum alloy by Nd:YAG pulsed laser", *Materials and Design*, 31(1), pp. 458–465.

[20] Casalino, G., Guglielmi, P., Lorusso, V. D., Mortello, M., Peyre, P., et al. "Laser offset welding of AZ31B magnesium alloy to 316 stainless steel" *Journal of Materials Processing Technology*, 242, pp. 49–59.

[21] Zuo, D., Hu, S., Shen, J., Xue, Z., 2014. "Intermediate layer characterization and fracture behavior of laser-welded copper/aluminum metal joints" *Materials and Design*, 58, pp. 357–362.

[22] Hongbo, X., Wang Tao, Liqun Li, Caiwang Tan, Kaiping Zhang, Ninshu Ma, 2020. "Effect of laser beam models on laser welding–brazing Al to steel" *Optics & Laser Technology*, pp. 122.

[23] Nasiri, A. M., Weckman, D. C., Zhouw, Y., 2015 "Interfacial microstructure of laser brazed AZ31B magnesium to Snplated steel sheet," *Welding Journal*, 94, pp. 61–72, 2015.

[24] "Z-Tech Lasers," Z-TECH advanced Technologies, 2017. [Online]. Available: http://ztechlasers.com/laser-welding-applications/

[25] "Application of laser beam welding," Mechanical Education, 2023. [Online]. Available: www.mechanicaleducation.com/application-of-laser-beam-welding/

[26] "6 Application of laser beam welding in industry", [Online].Available: www.xometry.com/resources/sheet/laser-welding-applications/

[27] Perveen, A, Molardi, C., Fornaini, C., 2018. 'Applications of laser welding in dentistry: A state-of-the-art review', *Micromachines*, 9(5), pp. 209. https://doi.org/10.3390/mi9050209

[28] Bi, J., Wu, L., Li, S., Yang, Z., Jia, X., Starostenkov, M.D., Dong, G., 2023. Beam shaping technology and its application in metal laser additive manufacturing: A review. *Journal of Materials Research and Technology.*

[29] Scherillo, F., Astarita, A., Prisco, U., Contaldi, V., di Petta, P., Langella, A., Squillace, A., 2018. Friction stir welding of AlSi10Mg plates produced by selective laser melting. *Metallography, Microstructure, and Analysis,* 7(4), pp. 457–463.
[30] Kimura, T., Nakamoto, T., 2017. Thermal and mechanical properties of commercial-purity aluminum fabricated using selective laser melting. *Materials Transactions*, 58(5), pp. 799–805.
[31] Cui, L., Peng, Z., Chang, Y., He, D., Cao, Q., Guo, X., Zeng, Y., 2022. Porosity, microstructure and mechanical property of welded joints produced by different laser welding processes in selective laser melting AlSi10Mg alloys. *Optics & Laser Technology,* 150, pp. 107952.
[32] Abroug, F., Monnier, A., Arnaud, L., Balcaen, Y., Dalverny, O., 2022. High cycle fatigue strength of additively manufactured AISI 316L Stainless Steel parts joined by laser welding. *Engineering Fracture Mechanics,* 275, pp. 108865.

10 Hybrid Welding Processes: Challenges and Future Perspective

Brij Mohan Mundotiya

1. INTRODUCTION

Single-pass welding is typically preferred for joining metallic parts. Different methods are available for achieving single-pass welding for thin sheets. However, single-pass welding cannot be used for thicker sheets when employing standard arc welding techniques. For thicker sheet welding, high-power electron beam, laser beam, hybrid laser-arc welding, and hybrid laser-friction stir welding are commonly used. These welding techniques can achieve high weld speeds and significant penetration depths. In addition, it shows high performance and speed, excellent flexibility, minimal distortion, and so on. Nevertheless, it has certain drawbacks, such as the expensive equipment cost and the difficulty of melting highly reflective and thermally conductive metals. In contrast, arc welding is popular because of its low-cost equipment and simple operation. However, it has poor weld penetration, slow welding speed, and easy hump development at high welding speed.

Hybrid laser-arc welding and hybrid laser-friction stir welding have emerged as potential techniques due to their ability to overcome the limitations of arc welding, friction stir welding, and laser welding. The research on hybrid laser-arc welding began in 1978, with Prof. Steen and colleagues publishing the first article on TIG-enhanced laser welding. However, this welding did not find rapid use in the industry due to the lack of commercial laser technology at the time. In the early 1990s, a multi-kilowatt laser system emerged that transferred beam source performance problems toward the fit-up and component tolerance. In 2000, Fraunhofer ILT successfully integrated the first hybrid laser-MIG welding system within a German oil tank manufacturing facility. Since then, the rapid growth of hybrid welding has been observed, and many new implementations in several industries, including the shipbuilding industry [1, 2] and the auto industry [3].

High-power laser sources with improved beam quality have recently been developed, including disc and fiber lasers. These sources have even higher high-power densities. This allows for deep penetrating welds with high aspect ratios in

DOI: 10.1201/9781003435884-10

thick-plate welding but also raises the requirements for edge quality. These fit-up issues could be solved by hybrid welding.

2. HYBRID LASER-ARC WELDING

Welding integrates a concentrated laser beam with an arc to generate and manipulate a weld pool during welding. The arc imparts heat to the upper weld region, which results in the distinctive "wineglass shape" of the weld seam [4]. Hybrid welding involves continuously feeding a metal wire, which functions as an electrode, from a spool. An electric arc results from connecting the workpiece and electrode to a power source. This electric arc generates a high heat energy, which causes the electrode and parts of the workpiece to melt and finally form a good weld joint. An inert gas, most frequently argon, is delivered to the welding area to provide a protective atmosphere. In addition, a laser beam is concentrated in the same region to create a deeper weld. When considering the various operations, it can be observed that using an arc and a focused laser beam offers a wider range of parameters.

Nowadays, the most widely used arc source in hybrid welding is MIG, MAG, TIG, and, less frequently, plasma arc welding (PAW). Among these, the MIG arc power emerges as the advantageous option for hybrid welding. MIG is preferred because of the inherent filler wire feeding into the weld pool and the easy procedure. The procedure can be tailored so that the MIG accurately dispenses the liquefied filler material into the gap. Simultaneously, the laser creates a vapor capillary inside the weld pool, ensuring deep penetration at high speed. The application of hybrid welding is continuously increasing due to its gap-bridging capabilities. From an industrial standpoint, the gap-bridging effect is crucial because it may offer the false impression that a good weld has been made when viewed from the top [5].

Determining the position between the arc and the laser beam (D_{LA}) is essential in hybrid welding. When the D_{LA} exceeds the approximate range of 5–8 mm, the resulting welding technique is called tandem or combination welding. The lack of interaction between them has potential advantages and drawbacks. Conversely, when the D_{LA} is equivalent to or smaller than the radius of the arc plasma, the resulting welding technique is referred to as coupled or hybrid welding. The main variation between them is that hybrid welding shows a synergy effect and consists of a hybrid weld pool. Another process variation is the arc position. The arc position significantly affects the weld quality. The leading laser/trailing arc and the trailing laser/leading arc are the two variations in the position of the arc. Generally, the leading laser/trailing arc configuration offers a wider upper bead part because of the flow of molten filler wire and arc forces. In contrast, the trailing laser/leading arc configuration, especially when using CO_2 lasers, provides greater penetration depth because of its preheating mechanism [6].

In hybrid welding, argon is the predominantly used gas in the welding process. Incorporation of the oxygen gas in argon gas promotes the detachment of droplets and reduction of spattering. The addition of helium in this gas mixture raises the arc voltage, which results in an increase in power and obtains wider seams. However, there is a reduction in arc stability [7].

2.1. Interaction between laser beam and arc

The interaction phenomenon is affected by several factors, including the type of laser and shielding gas, arc current, and electrode angle. These factors affect the heat transfer capabilities of the heat source. Abilash et al. [8] studied the influence of welding direction on leading laser/trailing arc configured hybrid welding of carbon steel. The welding direction influenced the laser-arc synergic effects. This synergy effect altered the melt flow, which improved the joint quality. They also reported that the synergy effect influenced the melting energy of the workpiece.

Typically, there are two components to welding energy. A portion of this energy is lost through conduction, convection, and heat radiation, while the workpiece and filler materials absorb the remaining part. The heat radiation and spatters from the molten pool cause some of this received energy to be wasted. The workpiece and filler materials are melted using the remaining energy, which is referred to as melting energy. Welding energy is the most important parameter since it determines welding efficiency. The welding efficiency will be increased by increasing the energy utilization. Energy utilization refers to the ratio of utilized energy to the total available energy to melt welding materials. It is mathematically expressed as Δ_U at constant welding speed [9]:

$$\Delta_U = \frac{E_H - (E_L + E_A)}{E_L + E_A} = \frac{A_H - (A_L + A_A)}{A_L + A_A} \qquad \text{Eq. (1)}$$

Where E and A represent the melting energy and cross-sectional area, respectively. The subscripts H, L, and A belong to hybrid, laser, and arc welding, respectively. The A_H and $A_{L,}$ and Δ_U increase as laser power increases. When considering arc current, the values of A_H and A_L increase as the arc current increases, while the change in Δ_U occurs contrary. The synergy effect is also determined through spectral intensity ψ_I [9]. "ψ_I" refers to the spectral intensity ratio of laser-arc hybrid welding to single heat source welding. The following is the expression for ψ_I:

$$\Psi_I = \frac{I_H - (I_L + I_A)}{I_L + I_A} \qquad \text{Eq. (2)}$$

Where I indicates the spectral intensity. The higher the ψ_I value, the larger the synergy effect. From Figure 10.1, the value of ψ_I increased with increasing laser power. However, it was decreased by an increase in arc current and D_{LA}. These results indicated that the synergy effect increased by raising the laser power. Furthermore, it can be noted that the Δ_U exhibits an increase in correlation with the rise in the variable ψ_I. The maximum value of the Δ_U is 0.2846, typically larger than 0.

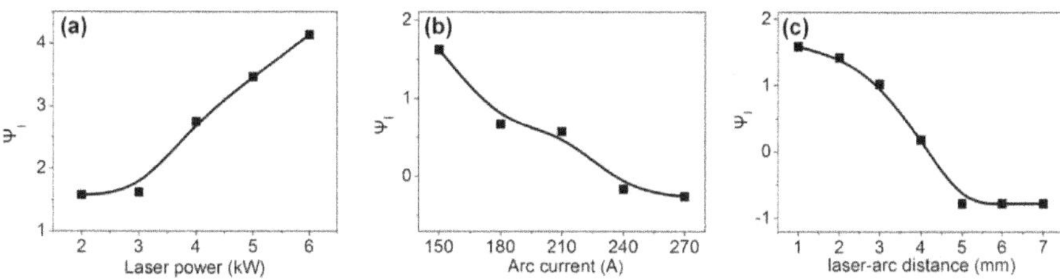

FIGURE 10.1 Laser-arc synergic effect when: (a) laser power (I = 150 A, D_{LA} = 2 mm); (b) arc current (P = 3 kW, D_{LA} = 2 mm); (c) laser-arc distance (P = 3 kW, I = 150 A) [9]

Meng et al. [10] introduced a new parameter called the melting energy increment (Ψ) to measure the synergy effect quantitatively. The expression for ψ is as follows:

$$\Psi = \frac{S_H - (S_L + S_A)}{S_L + S_A} \times 100\ \% \qquad \text{Eq. (3)}$$

Where S indicates the cross-sectional area. The results revealed that the melting efficiency and synergistic effect increase with increasing Ψ. The value of Ψ depends on the D_{LA}. It was changed when the D_{LA} rose from 1 to 4 mm, but its value remained unchanged after it reached 4 mm and above. It indicated that the laser-arc plasma interaction decreased when the D_{LA} was enhanced, reducing the synergy effect [11, 12].

The Ψ of hybrid laser-TIG welding exhibited a substantial increase from 27.4% to 83.6% as the D_{LA} was changed from 1 to 3 mm. However, it experienced an abrupt decline to 55.8% when the D_{LA} was further increased to 4 mm [10]. The reduction in the Ψ indicated that a higher-intensity laser beam could more easily irradiate the filler wire and droplet.

2.2. Hybrid laser-arc welding process parameters for proper joint quality

Hybrid welding is deemed suitable for a wide range of industrial applications. Nevertheless, appropriately configuring various welding parameters is necessary to achieve desirable joint quality. Further elaboration will be provided on certain parameters in subsequent sections.

2.2.1. Energy input

The energy input (or heat input) influences the microstructure, heat-affected zone (HAZ) width, penetration depth, and joint mechanical properties. Wang et al. [13] examined the impact of different heat inputs on the weld geometry of 800 MPa hot-rolled Nb-Ti-Mo microalloyed steel. From Figure 10.2, the width of the fusion zone (FZ) and HAZ expanded, and penetration depth diminished from 1.76 to 0.99 by increased heat input. A change in the geometric shape of the FZ was obtained. The shape was transformed into a “wine cup-like” shape. Figure 10.2a demonstrated

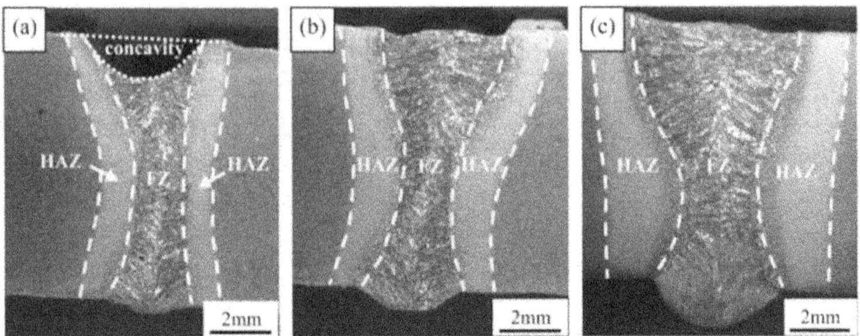

FIGURE 10.2 Surface morphologies in cross-section of weld junctions at heat inputs: (a) 3.90 kJ/cm, (b) 5.20 kJ/cm, and (c) 7.75 kJ/cm [13]

that the fusion zone consisted of a concavity by enhancing the welding speeds and decreasing the heat input. Üstünda et al. [14] also reported similar behavior. The upper portion of FZ had a larger width than the root due to the presence of the V groove.

Yu et al. [15] explored a study on hybrid welding. In the welding process, 16-mm thick TC4 titanium alloys were employed. The results revealed that when the laser power was raised to 4200 W, the penetration depth went up to 2.5 mm. Moreover, laser power increased filler metal height. The evaporation of the high-temperature liquid metal was highly intensive at higher laser powers, resulting in a larger clockwise vortex formation. This phenomenon induced the aggregation of molten metal and caused the filler metal height to increase. Leo et al. [16] investigated the impact of arc and laser power on the welds of an Al–Mg alloy. Laser power decreased Mg content and increased porosity in FZ. The porosity and the change in Mg content can influence the welded joint's tensile strength and ductility. Segregated coarse phases were also present in the FZ. The porosity and segregated coarse phases caused a stress concentration; thus, the cracks will easily originate from that location and spread across the section. In addition, the crack tip mobility was enhanced due to the larger grain size observed in the FZ. As a result, the tensile strength and ductility were both dropped. Yan et al. [17] reported that the weld quality of AA6005-T5 aluminum alloy joints was influenced by heat input. The FZ and HAZ of the welded part did not consist of liquidation or solidification cracks. Lei et al. [18] observed the presence of cracks in the weld crater at the weld bead end in hybrid laser-MAG welded joints of 30CrMnSiA. The cracks were solidification cracks that occurred at high temperatures. The energy input also impacted the residual stress in welded joints.

2.2.2. Welding speed

It is a critical processing parameter that reduces welding stability. It influences the weld width, weld penetration, and arc stability. Deeper penetration is achieved by slowing down the welding speed, however, this relationship is non-linear, especially at lower welding speeds. This issue is addressed by increasing the laser power and filler wire feed rate [19]. An increased wire feed rate improves arc rooting, which solves these difficulties.

Cao et al. [20] investigated that the microstructure as well as the mechanical properties of welded butt joints made of different metals, such as AISI316L and EH36, were affected by the welding speed. A welding speed below 0.8 m/min caused higher penetration, a primary defect in the welded joint. An optimal seam appearance was achieved at ~ 0.8 m/min. However, welding speeds over 1.0 m/min produced incomplete penetration and porosity. Cao et al. [21] reported the influence of welding speed on the penetration depth and defect formation in high-strength low-alloy steel welds. Turichin et al. [22] carried out hybrid welding experiments using X80 steel of 14 mm thickness and explored mechanical properties that changed with changing the welding speed. Higher welding speeds resulted in faster cooling rates and a higher percentage of martensite, while lower welding speeds produced the reverse trend. In addition, higher welding speed provided a faster heating and cooling cycle, creating a steeper temperature gradient at the weld zone. This temperature gradient can lead to metallurgical defects inside the weldment [23]. According to Kim et al. [24], inadequate fusion and insufficient weld penetration were obtained when the welding speed was excessively high. In contrast, a welding speed that was too low resulted in a bigger weld pool with deep penetration that extended beyond the point at which the material could be burned, resulting in partial disintegration.

The penetration depth can be enhanced by raising the root size. At a root size of 2 mm, the weld completely penetrated. However, if the root size is increased above a specific limit, the welded joint will have poor surface quality, root humping, and sag defects [25]. These defects are caused by gravity. Ceramic backing support may prevent sagging defects from developing at the root surface.

2.2.3. Relative position of the laser and the arc torch along the welding direction

Weld characteristics mainly depend on whether the arc torch is leading or trailing. The arc-leading (HM) arrangement contributes to achieving a higher degree of penetration. According to Tang et al. [26], the HM configuration produced superior melt dynamics, which corresponded to reduced melt velocities in the root region. A leading laser (HL) configuration with an air gap is not recommended because of the inadequacy of the material for creating the keyhole. Although, Casalino et al. [27] reported that the HL configuration resulted in a significantly deeper weld penetration. Hirohata et al. [28] used HM configuration and reported satisfactory joints on 15-mm thick plates. Two different heat source configurations were examined by Zhang et al. [29] to determine how the arc morphology and droplet transfers changed over time. The HM configuration had better results when comparing the morphologies of the weld beads produced by the two processes' configurations. The impact of heat source arrangement on weld formation was assessed by Kah et al. [30]. The study concluded that the HM configuration resulted in a comparatively broad and smooth surface, thereby ensuring the efficient functioning of the gas shielding mechanism. The impact of HM configuration on weld root drop-out was investigated by Tang et al. [26]. In the HM configuration, microscopic droplets caused by the gradual accumulation of molten pools successfully suppressed the production of humping. This was in contrast to the process of HL configuration. On the contrary, the utilization of the HL configuration yielded a comparatively more stable arc in comparison to the HM configuration because the stability of the keyhole was high in the former. The weld penetration was improved by 10% when the MAG torch was positioned at the rear of the CO_2 laser beam [31].

Furthermore, it is evident from the findings that the D_{LA} is a crucial component that controls the welding state throughout the welding procedure. It is discovered that adjusting the heat source status using the parameter D_{LA} improves the molten pool's stability. Campana et al. [32] conducted a hybrid welding experiment for joining 8-mm-thick plates. Their findings indicated that the maximum penetration depth arrived when D_{LA} was between 2 and 4 mm. Conversely, it decreased when the D_{LA} was above 6 mm. The reproducibility of the results was severely impaired when the D_{LA} was zero. Liu et al. [33] also studied the influence of D_{LA} on the penetration depth of various hybrid welding configurations. In TIG welding, specifically in the HM configuration, the penetration depth initially rose and then fell by increasing the D_{LA} from 1 to 7 mm. Indeed, an increase in the D_{LA} consistently decreased penetration depth in the HL configuration. According to Zhang et al.'s [29] analysis, it was determined that the successful production of high-quality welds was achieved at D_{LA} of 2–4 mm. This outcome was attributed to the significant synergistic interaction between the two leading configurations. There was an occurrence of the spatters when the D_{LA} was small. When D_{LA} exhibited high value, it formed porosity and undercut in the HM configuration, while spatters were observed in the HL configuration. Liu et al. [34] performed hybrid welding experiments using a 2-mm thick titanium alloy plate. Their findings revealed that D_{LA} affected welding stability. A full penetration on the rear side of the seam was achieved when D_{LA} was 1 mm. Furthermore, the surface exhibited a state of continuity, displaying a consistent and even appearance on both its upper and rear sides. A burn-through defect ran continuously along the weld seam when the D_{LA} was increased to 2 mm. Further, an increase of D_{LA} over 4 mm resulted in an insufficient fusion at the rear side of the seam.

2.2.4. Focal point position

The focal point position affects joint stability, quality, and penetration. In hybrid welding, the optimal weld penetration can be obtained by the position of the laser beam beneath the welded sheet. Campana et al. [32] performed hybrid welding experiments on 8-mm thick plates. Their findings revealed that optimal results were obtained by focusing the focal point of the laser beam within the joint. Unt and Salminen [35] investigated the impact of the focal point position (above and within the joint) on the penetration depth and shape weld in structural steel joints. The weld achieved with a focal point position (fpp) above +2 mm exhibited a more pronounced arc energy input than the weld obtained with an fpp below -2 mm. However, the melt area was high for the fpp -2 mm, indicating the welding process had high energy. The shape of the top bead was a narrow weld with a convex face for fpp +2 mm. Conversely, a wider weld with a concave face was yielded for fpp -2 mm. The focal point beneath the specimen surface was preferred, according to Zhang et al. [36]. The weld depths that can be achieved with a negative defocus were much greater than those that can be achieved with either a zero or a positive defocus. This phenomenon occurred due to the fact that when the laser was defocused negatively, the power density within the keyhole wall was higher compared to situations where the laser beam diverges outward at either zero or positive defocus. The ability of laser energy to penetrate deeper into a material was enhanced when a deeper keyhole was formed, as this facilitated multiple reflections on the walls of the keyhole [37, 38].

2.2.5. Angle of electrode

The conventional approach involves selecting an electrode angle ranging from 40 to 70° relative to the workpiece surface [39-44]. This phenomenon contributed to the reduction in the magnitude of the arc. It is a common practice to use a perpendicular laser beam on the workpiece to increase penetration. However, a tilted laser beam is applied to the workpiece if the workpiece has high reflectivity. The high-reflecting workpiece will reflect the laser beam toward the laser equipment, which damages it. To prevent damage, it is common practice to maintain a fixed angle between 5–10° for a tilted laser beam [17, 21, 44, 45]. The orientation of the electrode significantly impacts the appearance of the welds formed during the welding.

A curved weld seam became apparent when the electrode angle exceeded 55°, as the apex of the filler wire was noticeably distant from the surface of the workpiece. The directional property of the feeding wire experiences a decline, resulting in increased difficulty for the droplet to move quickly. Moreover, the unstable liquid droplet will induce an unpredictable trajectory, forming a curved fusion joint. However, a satisfactory fusion joint forms when the electrode angle is below 55° [46]. It can be attributed to a homogeneous and stable transition of droplets into the weld pool.

2.2.6. Joint gap (or gap-bridging)

From an industrial perspective, gap-bridging is critical because it may give a false perception of a satisfactory weld when seen from the top. Gap-bridging determines weld bead and root form. When the gap between joints is widened, there is a reduction in the base metal that undergoes melting, resulting in a decrease in dilution. If the gap widens further, the melt will eventually fall into it. A 0.2 to 0.25 mm gap can be used to weld components with the laser beam. However, for parallel sides, the gap width restrictions are typically between 1 and 1.5 mm [47-49].

In a study by Lamas et al. [50], the upper reinforcement initially exhibited an undercutting effect, which subsequently transitioned into underfilling over time when a widening gap existed. Figure 10.3 shows a change in the cross-sectional view of the hybrid welds with a large gap. The weld geometry transformed from a "cup-shaped" form into a "vase-shaped" form [22]. According to Wang et al. [51], a complete joint penetration occurred at a low gap size. Partial joint penetration was obtained with the widening of the gap size. The surface collapsed due to insufficient filler weld metal at a large gap size.

The joint gap also influences welding efficiency. The observed trend indicates a decrease in welding efficiency as the joint gap increases. The reduction in efficiency was observed due to a large amount of melted metal needed to fill the large gap in the welded joint. Turichin et al. [22] conducted hybrid welding experiments on a 14-mm-thick X80 steel. Their findings revealed that the maximum efficiency was achieved at a joint gap of 0 mm. Reisgen et al. [52] reported a positive correlation between the wire feed rate and the joint gap. The wire feed rate increased with an increase in the

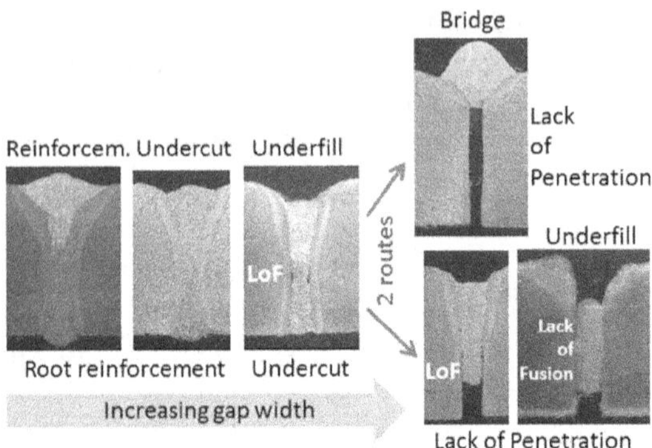

FIGURE 10.3 Effect of gap width on weld joint quality [50]

joint gap. Additionally, the penetration depth increased with joint gap width, reaching a maximum at a 1 mm gap. The penetration began to decrease above this joint gap.

2.3. Root hump formation

Hybrid laser-arc welding has been observed to exhibit an issue of excessive root penetration when attempting single-pass full penetration on sheets with a thickness exceeding 10–12 mm. This defect is classified as continuous root sagging or intermittent droplet production, also known as root humping [36, 53, 54]. The porosity and lack of fusion are also related to root humping, which reduces the fatigue durability of the welded component. Excessive root penetration, a defect commonly observed in welding, is closely linked to weld cap underfill. According to Frostervarg [55], hybrid laser-arc welding frequently results in humping. The hump was formed because the weld pool had poor melt dynamics and an increased root width. Berger et al. [56] explored the phenomenon of humping in welds with wide/shallow and narrow/deep configurations. They observed that the rapid solidification of the melt pool at high speed restricted the melt flow, leading to its swelling. The melt pool expanded further when the heat source moved away from the formed hump. However, the expansion of the melt pool will once again be choked off by the solidification of the melt pool, resulting in subsequent swelling occurring near a heat source. Furthermore, it found that the formation of humps occurred when melt stream velocity was larger than the welding speed. Blecher et al. [57] also noticed root humping during the laser-arc welding procedure. They suggested that the root hump formed due to high heat inputs. Ohnishi et al. [58] examined the impact of the shielding gas on the welding process and found that an excessively efficient shielding gas led to a decrease in oxygen content within the molten material. Consequently, this reduction in oxygen content resulted in decreased viscosity, less penetration, and the creation of a hump. Bunaziv et al. [25] reported that root humping was observed in laser-arc hybrid welding due to adverse melt flow at the bottom of the weld pool. In their experimental results, the root-humping-free welds below 1.2 m/min welding speed and 14 kW laser beam power.

2.4. Microstructure-mechanical property relationship

The welded joints consisted of three regions: (a) the fusion zone (FZ), (b) the HAZ, and (c) base metal (BM). Due to high solidification rates, most weld joints consist of the dendritic microstructure near the FZ. Hao et al. [4] established a toughness-microstructure relation after conducting hybrid welding experiments on a 4-mm-thick sheet of martensitic stainless steel. The HAZ of these welds consisted of the three regions starting from the FZ to toward BM: (a) coarse-grain HAZ (CGHAZ), (b) fine-grained HAZ (FGHAZ), and inter-critical HAZ (ICHAZ). A similar kind of weld-shape geometry with different heat-affected zones was reported by Üstündağ et al. [14]. A large CGHAZ produced a large grain boundary region across which stress relaxation could be accommodated, which decreased the material's sensitivity to hot cracking. The grain coarsening was anticipated to be a reduction in strength. Üstünda et al. [14] examined hybrid laser-arc welding to study X120 steel weld microstructure. The grain size in CGHAZ and FGHAZ was increased with enhancement in the heat input. The FGHAZ exhibited a reduced exposure to higher peak temperatures as a result of its large distance from the FZ and heat sources. Consequently, this region was predominantly composed of fine grains, leading to the expectation of enhanced impact toughness in these regions of the weld joint.

Skowronska et al. [59] employed a 10-mm-thick thermomechanically treated S700MC steel to conduct hybrid plasma-MAG welding. The results revealed that the welds exhibited a dendritic nature with a characteristic bainitic-ferritic microstructure. Additionally, the HAZ consisted of two distinct sub-zones characterized by varying grain sizes. Due to variations in grain sizes, the HAZ demonstrated a different range of hardness. Yan et al. [17] examined the impact of heat input on the microhardness and strength of joints in AA6005-T5 aluminum alloy. In the FZ, the alloying elements were vaporized. Therefore, it had a lower hardness value than the HAZ [60, 61]. The microhardness of the HAZ exhibited a lower value compared to the BM in the alloy connections. The decrease in hardness was attributed to a lower quantity of larger precipitates in the HAZ compared to the BM. The fraction and size of the precipitates also affected the strength of the weld. When the size of the precipitates exceeds 3–5 nm, dislocations tend to avoid them, forming a closed dislocation loop. This phenomenon is commonly referred to as the Orowan mechanism. The correlation between strength, S, and precipitates can be formulated as follows:

$$S \propto \sqrt{f_v}/r \qquad \text{Eq. (6)}$$

In Eq. (6), fv denotes the volume fraction of precipitation, and r is the average radius of precipitation. The f_v value of AA6005-T5 aluminum alloy joints decreased as the r increased within the FZ. Consequently, there was a decrease in strength. Bunaziv et al. [62] examined the effect of heat input on the welding of thick HSLA steel joints. In the root region, the cooling rate was reduced with increased heat input. The root microstructure was softer and more ductile. However, the reduction in the cooling rate resulted in a large-width CGHAZ. Cao et al. [20] investigated that the microstructure as well as the mechanical properties of welded butt joints made of different metals, such as AISI316L and EH36, were affected by the welding speed. The results revealed that martensite constituted the predominant phase within the central region of the seam. The observed phenomenon in the martensite zone can be determined by the ratio of Ni_{eq} to Cr_{eq}. The microhardness of the joint exhibited considerable variation across the weld section, measuring approximately 350 HV0.3. This value was twice as high as the microhardness of the base materials, which obtained 180 HV0.3.

3. HYBRID LASER-FRICTION STIR WELDING

Friction stir welding (FSW) is a prospective joining method that uses a spinning tool to join metal parts without melting metals that have differing physical properties. Low melting point materials, like aluminum, and materials that are challenging to weld using conventional processes, are frequently joined using this welding procedure. High tool wear and material requirements make the FSW technique challenging when joining high-temperature metals and alloys like steel and titanium. The tool wears out due to the increased load between it and the workpiece. In addition, lower welding speeds are used in the FSW due to high wear during the tool and workpiece plunging stage.

The viable approach to address these issues is a hybrid laser-FSW technique. A hybrid FSW process is defined as the inclusion of an extra laser heat source and FSW tool. During the welding process, the laser source preheats and softens the workpiece. Preheating the workpiece is beneficial for raising the temperature near the tool pin when joining high-temperature or dissimilar metals. It leads to easier welding of workpieces and less tool wear. In hybrid welding, the use of a laser source lowers the load between the tool and the workpiece. As a consequence, a high welding speed may be utilized to join the workpiece. According to Song et al.'s [63] research, laser as a preheating source in hybrid welding significantly enhances the mechanical properties of the Inconel 600 weld joint and refines the grain structure. The mechanical properties improved due to grain refining. Grain refining in hybrid laser-FSW happened as a result of dynamic recrystallization. High strain rates and frictional heat are present during this welding between the tool and the workpiece. The peak temperature rose between 810 and 845 °C when the welding speed and strain rate rose. The high strain rate and temperature are necessary for the dynamic recrystallization of grains. Fei and Wu [64] also reported that hybrid welding has much finer grains than the traditional FSW. Additionally, Sun et al. [65] reported an improvement in efficiency and tool life while using a hybrid laser-FSW to weld S45C steel plates that were 3.2 mm thick. The FSW of the S45C steel plates causes the formation of the brittle martensite and bainite phase above 300 mm/min. However, the formation of the brittle martensite and bainite phases can be avoided in hybrid laser-FSW by laser preheating at less than 600 mm/min welding speed.

The joining of dissimilar metals is difficult with FSW welding and hybrid laser-FSW. The problem with this dissimilar welding is that the joint performance drastically degrades due to the development of a brittle intermetallic layer close to the interface between two metals. Chang et al. [66] reported that the brittle intermetallic phase of $Al_{12}Mg_{17}$ can be eliminated or replaced with more ductile phases such as

NiAl and Ni_2Mg by using a Ni foil between the AA6061-T6 Al alloy and AZ31 Mg alloy plates. The joint efficiency was also increased by 66% in 2 kW laser power. It is considered that the increase in the efficiency was due to the formation of NiAl and Ni_2Mg phases instead of brittle intermetallic compound $Al_{12}Mg_{17}$.

Systematic research on welding efficiency and tool life has not yet been detailed. The emergence of this hybrid welding is presently in the early stages.

4. HYBRID WELDING AREAS OF APPLICATIONS

In today's world, the majority of attention paid to this process is concentrated on specific fields of application, including oil and gas pipelines, mobile equipment, mobile transport, shipbuilding, and road transportation, as well as the energy industry.

The hybrid laser-arc welding technology is used in various automotive industries, including JFE steel and the axle production factory at Daimler in Germany. The welding speed at Daimler's axle production facility exhibits a potential enhancement of approximately 30% compared to traditional welding. In another automotive application, an aluminum side panel located in front of the driver's compartment, doors of the Volkswagen Phaeton, extruded sections, and castings are welded using a hybrid process. The door was equipped with a total of 48 hybrid laser seams, each measuring 3,570 mm in length. The predominant type of seams observed on the lap joint were fillet seams, with butt seams being less frequently observed. The Audi A8 car's lateral roof frame was equipped with weld seams measuring 4.5 meters long and fitted with multiple functional sheets [67]. Airbus is interested in hybrid welding in the aerospace and aviation industry after years of successfully employing laser welding to produce aircraft fuselage sections for years. Regarding titanium alloys, Airbus is quite interested in this hybrid welding method. Traditional welding techniques are typically thought to make it nearly impossible to produce high-quality welds in thin titanium alloys. However, hybrid welding can meet this obstacle with good outcomes.

The effectiveness of hybrid welding in shipbuilding has long been established for many reasons, including (a) an increase in welding site performance, (b) significantly reduced material and energy requirements compared to traditional technologies, (c) the provision of highly accurate seams with low levels of distortion, and (d) the ability of the shipbuilding yard to lower unit costs and increase revenue [68]. Initially, only a few European shipbuilders who specialize in passenger ships may use this welding procedure. The approach is not well liked by those who construct commercial ships. In 2010, the hybrid welding procedure in ships saw its first commercial application. Large crude carriers (VLCCs), container ships, and cryogenic carriers (LNG or LPG carriers) are the target markets for this technique.

Hybrid welding is applied to the butt joints of the comparatively thin steel plates used in the upper structures of the hull and walls of the engine room. Gas-metal arc welding is commonly employed in shipyards as a conventional method for longitudinal welding on both sides. Due to the enormous amount of energy that is input during this welding method, substantial thermal distortion is produced. However, hybrid welding offers the advantage of reduced heat input and minimized distortion. The construction of onshore and offshore pipelines for oil and gas transportation is progressively gaining significance within the oil and gas sectors. Hybrid laser-MAG welding is a commonly employed technique to join pipelines. Adopting hybrid laser-MAG welding has received recognition for its application in an industrial environment. This welding technique has been implemented and evaluated compared to the conventional methods of three-pass MAG welding and laser welding with filler material incorporated.

Hybrid laser-arc welding is also commonly employed in mobile equipment like cranes, mining equipment, and agricultural machinery. Extreme durability and a high strength-to-weight ratio are requirements for these items. The widespread welding of the ultra-high-performance steels used in mobile equipment is made possible by hybrid welding. This welding technique is utilized extensively in the energy sector to manufacture generator turbines, wind towers, utility towers, and other essential energy components. Holtec International (Holtec) fabricates nuclear vessels, small modular reactors, and spent nuclear fuel dry storage canisters using hybrid laser-arc welding. This joining method resolves significant manufacturing issues related to the fabrication of nuclear components, such as narrow gap welding of the small modular reactor, reducing stress corrosion cracking in austenitic steels, and improvements in reliability, quality, and cost compared to conventional multi-pass manual welding.

5. SUMMARY

The hybrid welding process now promises to be very productive. Some benefits include faster welding speeds, the ability to bridge gaps, a lower thermal load, increased thickness, improved weld quality, and reduced operating costs. The welding parameters, such as heat input, welding speed, laser positions along the welding path, focal point position, electrode angle, and joint gap, are all explained in this chapter. Furthermore, the relation between the quality of the joint and its mechanical properties with these parameters is thoroughly investigated. The above works by various research groups can draw the following findings. The synergy effect can be adjusted to manipulate the properties of the weld joints. The optimization of the synergy effect can be achieved through the manipulation of various welding process parameters. In addition to improving welding speed and depth of penetration, the synergy effect strengthens the capacity to bridge gaps and provide a steady welding process. In the case of the hybrid laser-FSW process, high welding efficiency may be achieved using a laser beam, which allows for fast welding speed while reducing tool wear.

However, owing to the substantial capital investment and the complicated process nature of many process parameters, this welding technique only sees a moderate expansion in today's industry.

REFERENCES

1. H. Lembeck, *Laser-Hybrid-Schweißen im Schiffbau, Proceedings Aachener Kolloquium für Lasertechnik*, Aachen, Germany, September, (2002) 177–192.
2. U. Jasnau, J. Hoffmann, P. Seyffarth, R. Reipa, and G. Milbradt, *LaserMSG-Hybridschweißen im Schiffbau, Proc. European Automotive Laser Application*, Bad Nauheim, Germany, January, (2003).

3. T. Graf, and H. Staufer, Laser-hybrid-welding drives VW improvements, *Welding Journal*, January, (2003) 42–48.
4. K. Hao, C. Zhang, X. Zeng, and M. Gao, Effect of heat input on weld microstructure and toughness of laser-arc hybrid welding of martensitic stainless steel, *Journal of Materials Processing Technology*, 245, (2017) 7–14. DOI: 10.1016/j.jmatprotec.2017.02.007
5. K. D. Lee, and K. Y. Park, A study on the process robustness of Nd:YAG laser-MIG hybrid welding of aluminum alloy 6061-T6, *Proceedings of the International Congress on Applications of Lasers & Electro-Optics*, 307, (2003), (2003). DOI: 10.2351/1.5060045
6. T. Hayashi, S. Katayama, N. Abe, and A. Omori, A. High-power CO_2 laser-MIG hybrid welding for increased gap tolerance. Hybrid weldability of thick steel plates with a square groove, *Welding International*, 18, (2004) 692–701. DOI: 10.1533/wint.2004.3318
7. D. Petring, Development in hybridization and combined laser beam welding technologies, *Handbook of Laser Welding Technologies*, Edited by S. Katayama, Woodhead Publishing, (2013) 478–502.
8. M. Abilash, K. Senthil, G. Padmanaban, and S. Thriumalini, The effect of welding direction in CO2 Laser-MIG hybrid welding of mild steel plates, *IOP Conference Series: Materials Science and Engineering*, 149, (2016) 012031. DOI: 10.1088/1757-899X/149/1/012031
9. C. Zhang, M. Gao, and X. Zeng, Influences of synergy effect between laser and arc on laser-arc hybrid welding of aluminum alloys, *Optics & Laser Technology*, 120, (2019) 105766. DOI: 10.1016/j.optlastec.2019.105766
10. Y. Meng, M. Gao, and X. Zeng, Effects of arc types on the laser-arc synergic effects of hybrid welding, *Optics Express*, 26, (2018) 14775–14785. DOI: 10.1364/OE.26.014775
11. M. Gao, X. Zeng, and Q. Hu, Effects of welding parameters on melting energy of CO_2 laser-GMA hybrid welding, *Science and Technology of Welding and Joining*, 11, (2013) 517–522. DOI: 10.1179/174329306X148138
12. K. Kang, Y. Kawahito, M. Gao, and X. Zeng, Effects of laser-arc distance on corrosion behavior of single-pass hybrid welded stainless clad steel plate, *Materials & Design*, 123, (2017) 80–88. DOI: 10.1016/j.matdes.2017.03.049
13. X. -N. Wang, S. -H. Zhang, J. Zhou, M. Zhang, C. -J. Chen, and R. D. K. Misra, Effect of heat input on microstructure and properties of hybrid fiber laser-arc weld joints of the 800 MPa hot-rolled Nb-Ti-Mo microalloyed steels, *Optics and Lasers in Engineering*, 91, (2017) 86–96. DOI: 10.1016/j.optlaseng.2016.11.010
14. Ö. Üstündağ, S. Gook, A. Gumenyuk, and M. Rethmeier, Hybrid laser arc welding of thick high-strength pipeline steels of grade X120 with adapted heat input, *Journal of Materials Processing Technology*, 275, (2020) 116358. DOI: 10.1016/j.jmatprotec.2019.116358
15. J. Yu, C. Cai, J. Xie, J. Huang, Y. Liu, and H. Chen, Weld formation, arc behavior, and droplet transfer in narrow-gap laser-arc hybrid welding of titanium alloy, *Journal of Manufacturing Processes*, 91, (2023) 44–52. DOI: 10.1016/j.jmapro.2023.02.022
16. P. Leo, G. Renna, G. Casalino, and A. G. Olabi, Effect of power distribution on the weld quality during hybrid laser welding of an Al–Mg alloy, *Optics & Laser Technology*, 73, (2015) 118–126. DOI: 10.1016/j.optlastec.2015.04.021
17. S. Yan, H. Chen, Z. Zhu, and G. Gou, Hybrid laser-metal inert gas welding of Al–Mg–Si alloy joints: Microstructure and mechanical properties, *Materials & Design*, 61, (2014) 160–167. DOI: 10.1016/j.matdes.2014.04.062

18. Z. Lei, B. Li, L. Ni, Y. Yang, S. Yang, and P. Hu, Mechanism of the crack formation and suppression in laser-MAG hybrid welded 30CrMnSiA joints, *Journal of Materials Processing Technology*, 239, (2017) 187–194. DOI: 10.1016/j.jmatprotec.2016.08.033
19. L. Bidi, S. Mattei, E. Cicala, H. Andrzejewski, P. L. Masson, and J. Schroeder, The use of exploratory experimental designs combined with thermal numerical modeling to obtain a predictive tool for hybrid laser/MIG welding and coating processes, *Optics & Laser Technology*, 43, (2011) 537–545. DOI: 10.1016/j.optlastec.2010.07.011
20. L. Cao, X. Shao, P. Jiang, Q. Zhou, Y. Rong, S. Geng, and G. Mi, Effects of welding speed on microstructure and mechanical property of fiber laser welded dissimilar butt joints between AISI316L and EH36, *Metals*, 7, (2017) 1–13. DOI: 10.3390/met7070270
21. X. Cao, P. Wanjara, J. Huang, C. Munro, and A. Nolting, Hybrid fiber laser — Arc welding of thick section high strength low alloy steel, *Materials & Design*, 32, (2011) 3399–3413. DOI: 10.1016/j.matdes.2011.02.002
22. G. Turichin, M. Kuznetsov, M. Sokolov, and A. Salminen, Hybrid laser arc welding of x80 steel: Influence of welding speed and preheating on the microstructure and mechanical properties, *Physics Procedia*, 78, (2015) 35–44. DOI: 10.1016/j.phpro.2015.11.015
23. L. J. Zhang, J. Ning, X. J. Zhang, G. F. Zhang, and J. X. Zhang, Single pass hybrid laser–MIG welding of 4-mm thick copper without preheating, *Materials & Design*, 74, (2015) 1–18. DOI: 10.1016/j.matdes.2015.02.027
24. D. Y. Kim, and Y. W. Park, Weldability evaluation and tensile strength estimation model for aluminum alloy lap joint welding using hybrid system with laser and scanner head, *Transactions of Nonferrous Metals Society of China*, 22, (2012) s596-s604. DOI: 10.1016/S1003-6326(12)61771-3
25. I. Bunaziv, C. Dørum, S. E. Nielsen, P. Suikkanen, X. Ren, B. Nyhus, M. Eriksson, and O. M. Akselsen, Laser-arc hybrid welding of 12- and 15-mm thick structural steel, *The International Journal of Advanced Manufacturing Technology*, 107, (2020) 2649 – 2669. DOI: 10.1007/s00170-020-05192-2
26. G. Tang, X. Zhao, R. Li, Y. Liang, Y. Jiang, and H. Chen, The effect of arc position on laser-arc hybrid welding of 12-mm-thick high strength bainitic steel, Optics & Laser Technology, 121, (2020) 105780. DOI: 10.1016/j.optlastec.2019.105780
27. G. Casalino, S. L. Campanelli, U. D. Maso, and A. D. Ludovico, Arc leading versus laser leading in the hybrid welding of aluminium alloy using a fiber laser, *Procedia CIRP*, 12, (2013) 151–156. DOI: 10.1016/j.procir.2013.09.027
28. M. Hirohata, G. Chen, K. Morioka, K. Hyoma, N. Matsumoto, and K. Inose, An investigation on laser-arc hybrid welding of one-pass full-penetration butt-joints for steel bridge members, *Welding in the World*, 66, (2022) 515–527. DOI: 10.1007/s40194-021-01221-0
29. S. Zhang, Y. Wang, M. Zhu, Y. Feng, P. Nie, and Z. Li, Effects of heat source arrangements on Laser-MAG hybrid welding characteristics and defect formation mechanism of 10CrNi3MoV steel, *Journal of Manufacturing Processes*, 58, (2020) 563–573. DOI: 10.1016/j.jmapro.2020.08.027
30. P. Kah, A. Salminen, and J. Martikainen, The effect of the relative location of laser beam with arc in different hybrid welding processes, Mechanika, 3 (83), (2008) 68–74. DOI: 10.2351/1.5061539
31. S. E. Nielsen, M. M. Andersen, J. K. Kristensen, and T. A. Jensen, Hybrid welding of thick section C/Mn steel and aluminium, *International Institute of Welding*, IIW doc. XII (2002) 1731–02.

32. G. Campana, A. Fortunato, A. Ascari, G. Tani, and L. Tomesani, The influence of arc transfer mode in hybrid laser-mig welding, Journal of Materials Processing Technology, 191, (2007) 111–113. DOI: 10.1016/j.jmatprotec.2007.03.001
33. L. M. Liu, S. T. Yuan, and C. B. Li, Effect of relative location of laser beam and TIG arc in different hybrid welding modes, *Science and Technology of Welding and Joining*, 17, (2012) 441–446. DOI: 10.1179/1362171812Y.0000000033
34. L. Liu, J. Shi, Z. Hou, and G. Song, Effect of distance between the heat sources on the molten pool stability and burn-through during the pulse laser-GTA hybrid welding process, *Journal of Manufacturing Processes*, 34, (2018) 697–705. DOI: 10.1016/j.jmapro.2018.06.038
35. A. Unt, and A. Salminen, Effect of welding parameters and the heat input on weld bead profile of laser welded T-joint in structural steel, *Journal of Laser Applications*, 27, (2015) S29002. DOI: 10.2351/1.4906378
36. M. Zhang, G. Chen, Y. Zhou, and S. Liao, Optimization of deep penetration laser welding of thick stainless steel with a 10 kW fiber laser, *Materials and Design*, 53, (2014) 568–576. DOI: 10.1016/j.matdes.2013.06.066
37. M. Vänskä, F. Abt, R. Weber, A. Salmine, and T. Graf, Effects of welding parameters onto keyhole geometry for partial penetration laser welding, *Physics Procedia*, 41, (2013) 199–208. DOI: 10.1016/j.phpro.2013.03.070
38. X. Jin, P. Berger, and T. Graf, Multiple reflections and Fresnel absorption in an actual 3D keyhole during deep penetration laser welding, *Journal of Physics D: Applied Physics*, 39, (2006) 4703–4712. DOI: 10.1088/0022-3727/39/21/030
39. L. Liu, X. Hao, and G. Song, A new laser-arc hybrid welding technique based on energy conservation, *Materials Transactions*, 47, (2006) 1611–1614. DOI: 10.2320/matertrans.47.1611
40. L. Liu, and R. Xu, Investigation of corrosion behavior of Mg-steel laser-TIG hybrid lap joints, *Corrosion Science*, 54, (2012) 212–218. DOI: 10.1016/j.corsci.2011.09.017
41. K. Zhang, Z. Lei, Y. Chen, M. Liu, and Y. Liu, Microstructure characteristics and mechanical properties of laser-TIG hybrid welded dissimilar joints of Ti–22Al–27Nb and TA15, Optics & Laser Technology, 73, (2015) 139–145. DOI: 10.1016/j.optlastec.2015.04.028
42. G. Casalino, A. Angelastro, P. Perulli, P. Posa, and P. R. Spena, Fiber laser-MAG hybrid welding of DP/AISI 316 and TWIP/AISI 316 dissimilar weld, *Procedia CIRP*, 79, (2019) 153–158. DOI: 10.1016/j.procir.2019.02.035
43. H. Gui, K. Zhang, D. Li, and Z. Li, Effect of relative position in low-power pulsed-laser–tungsten-inert-gas hybrid welding on laser-arc interaction, *Journal of Manufacturing Processes*, 36, (2018) 426–433. DOI: 10.1016/j.jmapro.2018.10.045
44. S. Yan, Y. Nie, Z. Zhu, H. Chen, G. Gou, J. Yu, and G. Wang, Characteristics of microstructure and fatigue resistance of hybrid fiber laser-MIG welded Al–Mg alloy joints, *Applied Surface Science*, 298, (2014) 12–18. DOI: 10.1016/j.apsusc.2013.12.157
45. C. Li, K. Muneharua, S. Takao, and H. Kouji, Fiber laser-GMA hybrid welding of commercially pure titanium, *Materials & Design*, 30, (2009) 109–114. DOI: 10.1016/j.matdes.2008.04.043
46. Z. Hou, C. Li, and L. Liu, Laser-TIG hybrid welding of magnesium alloy T-joint with cold filler wire, *Materials Transactions*, 56, (2015) 1242–1247. DOI: 10.2320/matertrans.M2015011
47. F. Vollertsen, and S. Gruenwald, Defects and process tolerances in welding of thick plates, *ICALEO 2008 - 27th International Congress on Applications of Lasers and Electro-Optics, Congress Proceedings*, (2008) 489–497. DOI: 10.2351/1.5061221

48. M. Rethmeier, S. Gook, M. Lammers, and A. Gumenyuk, Laser-hybrid welding of thick plates up to 32 mm using a 20 kW fiber laser, *Quarterly Journal of The Japan Welding Society*, 27, (2009) 74s-79s. DOI: 10.2207/qjjws.27.74s
49. Y. Yao, M. Wouters, J. Powell, K. Nilsson, and A. F. H. Kaplan, The influence of joint geometry and fit-up gaps on hybrid laser-MIG welding, *Journal of Laser Applications*, 18, (2006) 283–288. DOI: 10.2351/1.2355526
50. J. Lamas, J. Frostevarg, and A. F. H. Kaplan, Gap bridging for two modes of laser arc hybrid welding, *Journal of Materials Processing Technology*, 224, (2015) 73–79. DOI: 10.1016/j.jmatprotec.2015.04.022
51. H. Wang, Y. Wang, X. Li, W. Wang, and X. Yang, Influence of assembly gap size on the structure and properties of SUS301L stainless steel laser welded lap joint, *Materials (Basel)*, 14, (2021) 996. DOI: 10.3390/ma14040996
52. U. Reisgen, S. Olschok, and O. Engels, Laser beam submerged arc hybrid welding: A novel hybrid welding process, *Journal of Laser Applications*, 30, (2018) 042012. DOI: 10.2351/1.5037269
53. Q. Pan, M. Mizutani, Y. Kawahito, and S. Katayama, High power disk laser-metal active gas arc hybrid welding of thick high tensile strength steel plates, *Journal of Laser Applications*, 28, (2016) 12004.
54. T. Ilar, I. Erikson, J. Powell, and A. Kaplan, Root humping in laser welding–an investigation based on high speed imaging, *Physics Procedia*, 39, (2012) 27–32. DOI: 10.1016/j.phpro.2012.10.010
55. J. Frostevarg, Factors affecting weld root morphology in laser keyhole welding, *Optics and Lasers in Engineering*, 101, (2018) 89–98. DOI: 10.1016/j.optlaseng.2017.10.005
56. P. Berger, H. Hügel, A. Hess, R. Weber, and T. Graf, Understanding of humping based on conservation of volume flow, *Physics Procedia*, 12, (2011) 232–240. DOI: 10.1016/j.phpro.2011.03.030
57. J. J. Blecher, T. A. Palmer, and T. DebRoy, Mitigation of root defect in laser and hybrid laser arc welding, *Welding Journal*, 94, (2015) 73–82.
58. T. Ohnishi, Y. Kawahito, M. Mizutani, and S. Katayama, Butt welding of thick, high strength steel plate with a high power laser and hot wire to improve tolerance to gap variance and control weld metal oxygen content, Science and Technology of Welding and Joining, 18, (2013) 314–322. DOI: 10.1179/1362171813Y.0000000108
59. B. Skowronska, T. Chmielewski, D. Golanski, and J. Szulc, Weldability of S700MC steel welded with the hybrid plasma + MAG method, *Manufacturing Review*, 7, (2020) 1–15. DOI: 10.1051/mfreview/2020001
60. C. Liu, D. O. Northwood, and S. D. Bhole, Tensile fracture behavior in CO2 laser beam welds of 7075-T6 aluminum alloy, *Materials & Design*, 25, (2004) 573–577. DOI: 10.1016/j.matdes.2004.02.017
61. T. Y. Kuo, and H. C. Lin, Effects of pulse level of Nd–YAG laser on tensile properties and formability of laser weldments in automotive aluminum alloys, *Materials Science and Engineering: A*, 416, (2006) 281–289. DOI: 10.1016/j.msea.2005.10.041
62. I. Bunaziv, O. M. Akselsen, J. Forstevarg, and A. F. H. Kaplan, Deep penetration fiber laser-arc hybrid welding of thick HSLA steel, *Journal of Materials Processing Technology*, 256, (2018) 216–228. DOI: 10.1016/j.jmatprotec.2018.02.026
63. K. H. Song, T. Tsumura, and K. Nakata, Development of Microstructure and Mechanical Properties in Laser-FSW Hybrid Welded Inconel 600, *Materials Transactions*, 50, (2009) 1832–1837. DOI: 10.2320/matertrans.M2009058
64. X. Fei, and Z. Wu, Research of temperature and microstructure in friction stir welding of Q235 steel with laser-assisted heating, *Results in Physics*, 11, (2018) 1048–1051. DOI: 10.1016/j.rinp.2018.11.039

65. Y. F. Sun, Y. Konishi, M. Kamai, and H. Fujii, Microstructure and mechanical properties of S45C steel prepared by laser-assisted friction stir welding, Materials and Design, 47, (2013) 842–849. DOI: 10.1016/j.matdes.2012.12.078
66. W. -S. Chang, S. R. Rajesh, C. -K. Chun, and H. J. Kim, Microstructure and mechanical properties of hybrid laser-friction stir welding between AA6061-T6 al alloy and AZ31 Mg alloy, *Journal of Materials Science and Technology*, 27, (2011) 199–204. DOI: 10.1016/S1005-0302(11)60049-2
67. H. Staufer, Applications and case studies of laser hybrid welding in the automotive industry, *International Journal of Microstructure and Materials Properties*, 5, (2010) 234–244. DOI: 10.1504/IJMMP.2010.035942
68. M. Gogolukhina, L. Mamedova, O. Scholtz, and A. Firsova, Feasibility study of hybrid laser arc welding application in shipbuilding, *Key Engineering Materials*, 822, (2019) 459–466. DOI: 10.4028/www.scientific.net/KEM.822.459

11 Additive Manufacturing Integration with Welding

Laukik P. Raut, Ravindra V. Taiwade, Ashish Fande, Dhiraj Narayane, and Prachi Tawele

1. INTRODUCTION

1.1. Background

AM represents a paradigm shift in fabrication, allowing for the creation of complex structures layer by layer. Whether utilizing polymers, metals, or composites, AM has revolutionized prototyping, customization, and even end-use part production. Conversely, welding, a traditional yet indispensable technique involves joining materials through fusion, forming robust connections [1].

AM and welding, although distinct, share common goals of efficiency, precision, and material versatility. AM offers intricate design possibilities, while welding provides structural integrity. The convergence of these technologies holds immense promise, providing a holistic approach to manufacturing that marries the precision of AM with the durability of welding.

The motivation to integrate AM with welding is rooted in the complementary strengths of these techniques. By combining the design freedom of AM with the structural reliability of welding, manufacturers aim to unlock new frontiers in product development. This integration offers the potential to create components with intricate geometries, enhanced material properties, and improved overall performance. The pursuit of a synergistic relationship between AM and welding is driven by the quest for more efficient, flexible, and advanced manufacturing processes. This chapter delves into the motivations behind this integration, exploring how it addresses current limitations and opens doors to unprecedented possibilities in the realm of fabrication.

1.2. Objectives of Integration

AM revolutionizes design by offering unparalleled freedom. Traditional manufacturing methods often impose constraints on design due to tooling limitations, but AM liberates designers from these restrictions. The layer-by-layer construction allows for intricate geometries, internal structures, and customized features that were once impractical or impossible. This newfound design flexibility empowers engineers to create components optimized for function, weight, and performance, marking a paradigm shift in product development [2].

DOI: 10.1201/9781003435884-11

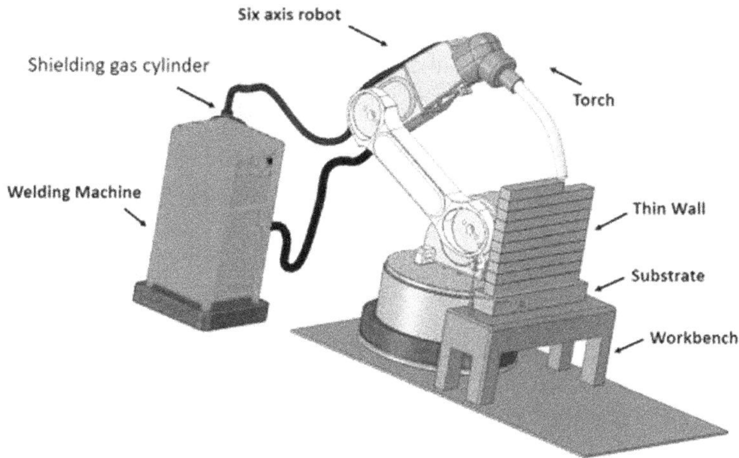

FIGURE 11.1 Schematic diagram of robotic wire arc additive [6]

The layer-wise construction inherent in AM contributes to enhanced manufacturing efficiency. Unlike subtractive methods that generate significant material waste, AM builds structures from the ground up, minimizing material usage. Additionally, the ability to consolidate multiple parts into a single, complex component reduces assembly requirements, streamlining the overall manufacturing process. The precision and repeatability of AM further contribute to efficiency gains, as intricate designs can be replicated with high accuracy, ensuring consistency across production batches.

AM opens avenues for exploring novel material combinations, pushing the boundaries of material science. The diverse range of materials compatible with AM processes includes polymers, metals, ceramics, and composites. Engineers can experiment with hybrid structures, combining materials with distinct properties to achieve desired mechanical, thermal, or electrical characteristics. This versatility enables the creation of tailored solutions for specific applications, from lightweight aerospace components to biomedical implants. The synergy between design flexibility and material exploration in AM fosters innovation in product development [1].

In the integration of AM with welding (Figure 11.1), these attributes of AM contribute to a comprehensive approach that leverages the unique strengths of both technologies. The following sections will explore how welding processes can be seamlessly incorporated into the AM workflow, addressing challenges and unlocking new possibilities in manufacturing.

2. OVERVIEW OF AM PROCESSES

AM comprises a diverse array of processes, each with its unique characteristics and applications. Understanding these processes is pivotal for comprehending the nuances of integrating AM with welding.

2.1. AM techniques

Selective laser sintering (SLS) involves the layer-by-layer fusion of powdered materials, typically polymers or metals, using a high-powered laser. The laser selectively sinters or fuses the powdered material, solidifying the layer and creating a three-dimensional object. SLS is celebrated for its versatility in working with various materials and its suitability for producing complex and functional prototypes, as well as end-use parts [3].

Stereolithography relies on a process where a laser solidifies liquid resin layer by layer. This technique is renowned for its precision and ability to produce intricate, high-resolution parts. SLA is often preferred for applications where fine details and smooth surface finishes are critical, making it a popular choice in industries like healthcare and aerospace.

Fused Deposition Modeling (FDM) is a widely used AM technique that involves the extrusion of thermoplastic filaments layer by layer to build up a structure. FDM is valued for its simplicity, cost-effectiveness, and accessibility. It is commonly used in rapid prototyping, concept modeling, and even in the production of functional parts, especially in industries like automotive and consumer goods [4].

Powder Bed Fusion (PBF) encompasses techniques such as Direct Metal Laser Sintering (DMLS) and Electron Beam Melting (EBM). In these methods, metal powders are selectively melted layer by layer to create fully dense metal parts. PBF is particularly prized for its ability to produce robust, high-strength metal components with intricate geometries, making it indispensable in aerospace, healthcare, and automotive applications.

Understanding the nuances of these AM processes is essential for evaluating their compatibility with welding techniques. The subsequent sections will explore how these AM techniques can be seamlessly integrated with welding processes, offering a comprehensive approach to advanced manufacturing.

2.2. AM materials and material compatibility with welding processes

Polymer-based materials are fundamental to additive manufacturing processes, offering a wide spectrum of options with diverse properties. Common polymers used in AM include ABS, PLA, nylon, and others. These materials are celebrated for their versatility, low cost, and ease of processing. Polymer-based AM is extensively employed in applications such as rapid prototyping, consumer goods, and medical devices [5].

Metal AM involves the use of various metals, including titanium, aluminum, stainless steel, and nickel alloys. Metal AM enables the production of high-strength, complex components suitable for industries such as aerospace, healthcare, and automotive. The ability to create intricate metal parts with additive manufacturing has transformed the landscape of metal fabrication [7].

Composite materials in AM combine polymers with reinforcing fibers or particles, providing a balance of strength, weight, and flexibility. Carbon fiber-reinforced polymers, for example, offer enhanced mechanical properties. Composites find applications in aerospace, sports equipment, and automotive industries, where the demand for lightweight yet robust materials is high [8].

Integrating AM with welding processes necessitates a thorough understanding of material compatibility. While AM allows for the fabrication of intricate structures with a variety of materials, welding introduces additional considerations. Matching the materials used in AM with those suitable for welding is crucial to ensure the structural integrity and performance of the final product. This compatibility extends not only to the base materials but also to any additional coatings or finishes applied during the AM or welding processes.

3. WELDING PROCESSES

3.1. Traditional welding techniques

Arc welding involves creating an electric arc between an electrode and the workpiece, generating the heat needed to melt and fuse metals. This versatile method is widely used in construction, fabrication, and repair. GTAW uses a tungsten electrode to create an arc for welding. It is commonly employed in applications where precision and control are paramount, such as in the aerospace and nuclear industries. GMAW, or MIG welding, utilizes a continuous wire electrode and shielding gas to join metals. It is favored for its speed and efficiency and is widely used in automotive, construction, and manufacturing.

3.1.1. GTAW-based WAAM

WAAM process which comprises the use of non-consumable tungsten electrodes along with a combination of shielding gases including argon, helium, hydrogen, carbon-dioxide, and so on, is called as gas tungsten arc welding (GTAW) or TIG welding. Metals fabricated using this type of WAAM include copper, magnesium, titanium, aluminum, and steel [9,10]. Out of the aforementioned metals and alloys, titanium, aluminum, and their respective alloys are of great importance to the aerospace industry.

3.1.2. PAW-based WAAM

Plasma Arc Welding can be understood as a development over GTAW with higher process freedom and better heat source. In GTAW the arc is created between the electrode and the substrate. In PAW the tungsten rod (cathode) is placed inside a hollow nozzle which is anode. The arc is struck between the rod and the nozzle through which the gas passes and gets ionized to form plasma which also acts as a heat source with the arc and has higher degree of process freedom as the gas plasma is not confined [11]. As the arc remains confined between nozzle and rod, it is called plasma non-transferred arc.

There exists another method called plasma-transferred arc where arc is not confined. Initially an arc is struck between the nozzle and the rod called pilot arc to form plasma with low current which helps to ignite the main arc "plasma transferred arc" between rod and the substrate for which the polarity of substrate is changed to positive and the pilot arc is subsequently extinguished [11]. As both arc and gas plasma acts as heat source we have both mobility and the temperature is roughly twice compared to GTAW. The twin-wire arc additive manufacturing process has been demonstrated to be capable of producing titanium aluminides at a reasonable cost. To address the issues about the previous gas tungsten arc welding-based process's low thermal efficiency, the twin-wire plasma arc additive manufacturing system is created in this study using compressed plasma arc power.

3.1.3. GMAW-based WAAM

GMAW is a consumable electrode type process. The electrode is same as the feedstock material which melts to get deposited, which results in efficient heat transfer and faster deposition [13,14]. But as the electrode melts it changes the arc length affecting the voltage and current which has to be then adjusted accordingly [15]. There are three modes of transfer in GMAW [11]. CMT and Tandem GMAW are two extended arms of GMAW [16].

3.1.4. Cold metal transfer (CMT)

In contrary to short circuiting here as soon as the partially molten droplet touches the surface the current is brought to 0 and the wire is retracted so no resistive heating and sputtering takes place. In order to detach the droplet, the current magnitude is set such that it won't detach on its own but will be able to deposit once it touches the substrate [11]. It brings along low equipment cost, high arc deposition rate and precision forming. It also has various variants like CMT, CMT-P, CMT-adv, CMT-PADV [17,18]. It reduces the post deposition working, increasing productivity and BTF ratio. Decrease in surface waviness was observed. The process can be optimized by optimizing the governing parameters via methods like GPR which has been further explored in the subsequent sections [19]. Shivraman found that CMT mode helped in improving microstructure and mechanical properties in aluminum alloys.

3.1.5. Tandem GMAW

It is another variant of GMAW. It is suitable to make thin structured walls. It employs two wires into molten pool simultaneously. It is believed to show higher efficiency and deposition rate up to (160 g/min). It requires higher wire feed rate and needs high heat input to maintain arc which may lead to molten pool overflowing due to excess heat accumulation, reduced heat dissipation, lowered viscosity, large arc force and strong droplet impingement [16,20]. To avoid pool overflow heat dissipation is the key.

3.2. Welding parameters

Welding Parameters primarily depend upon the thickness and height of the weld bead which is needed to be obtained. For obtaining ideal deposition parameters it is crucial that we carry out experiments so as to find their optimal relationship with layer height and cord width. With the inferences drawn from experiments carried out by J.L. Prado-Cerqueira et al, it was identified that there is a direct correlation between the wire feed rate and the intensity (measured in Amp (A)), and this correlation is independent of welding speed. There is ambiguity when it comes to the relationship between voltage and wire feed rate [21]. Theoretically, the voltage will lead to an increment of wire feed rate but in practice voltage rating is varying between 9V to 12V as a function of the wire speed. Another important process parameter is the intensity which has a considerable impact on bead geometry. When intensity (measured in amperes) is high and is in tandem with welding speed (more than 200 mm/min), then we observe smooth bead growth along with height as well as width. Welding speeds as high as 400 mm/min have shown excellent results including higher deposition and good penetration [22]. Another important welding parameter is heat input, where we focus on reducing the heat input and expecting to bring the excellent quality of fabricating parts in terms of macrostructure, microstructure, and mechanical properties. Heat input is also one of the variables used to assess heat-affected zone (HAZ) grain size and width [23].

3.3. Advanced welding techniques

Laser welding employs a high-intensity laser beam to melt and fuse metals. This precise and rapid technique is utilized in industries requiring fine welds, such as electronics and medical device manufacturing. In electron beam welding, a focused beam of high-velocity electrons is used to join metals in a vacuum. This technique is valued for its deep penetration and is employed in aerospace and automotive applications. Friction stir welding involves the use of a rotating tool to create heat and join materials without melting them fully. It is commonly used in joining lightweight materials, such as aluminum in the aerospace industry. Understanding these welding processes is pivotal for effectively integrating AM with welding, ensuring a seamless synergy between the two technologies. The subsequent sections will delve into the challenges and solutions associated with this integration, as well as explore real-world applications across various industries [24].

4. CHALLENGES AND SOLUTIONS

4.1. Managing thermal stresses during integration

The layer-by-layer nature of additive manufacturing (AM) can introduce thermal stresses during the build process. Variations in temperature between successive layers may lead to differential cooling and, consequently, the development of thermal stresses within the printed component. This phenomenon is particularly pronounced when combining AM with welding processes, as additional heat is introduced during the welding stage.

Optimized Printing Parameters: Fine-tuning AM parameters, such as layer thickness and print speed, can help mitigate thermal stresses. By optimizing these parameters, the rate of temperature change during the build process can be controlled, reducing the likelihood of stress accumulation [25].

Preheating Strategies: Preheating the substrate or incorporating controlled heating during the AM process can help maintain more uniform temperatures. This approach minimizes abrupt thermal transitions, mitigating thermal stress formation.

Sequential Printing and Cooling: Printing components in a sequential manner, with controlled cooling between layers, allows for gradual stress relief. This sequential approach reduces the build-up of internal stresses within the material [26].

4.2. Heat-affected zones and distortion control

Welding introduces localized heating, creating HAZ in the vicinity of the weld. The extent of the HAZ depends on factors such as welding parameters, material properties, and heat dissipation. Excessive HAZ can compromise the mechanical properties of the material and impact the structural integrity of the final component. Welding-induced heating can lead to distortion, altering the shape of the fabricated part. Controlling distortion is crucial, especially when integrating welding with the precise geometries achieved through AM.

Strategic Welding Sequences: Adopting specific welding sequences, such as alternating sides or using intermittent welding, can distribute heat more evenly. This minimizes the size and impact of the HAZ and helps control distortion [27].

Well-designed fixtures can constrain the part during welding, minimizing distortion. Fixtures act as heat sinks, controlling the cooling rate and reducing thermal gradients in the material. Applying controlled heat treatment after welding can relieve residual stresses and reduce distortion. PWHT allows for the adjustment of microstructural properties, enhancing the overall stability of the welded structure. By addressing thermal stresses and distortion control in both the AM and welding processes, the integration of these technologies becomes more robust. The following sections will delve into the compatibility of materials used in both processes and explore real-world applications across diverse industries [28].

4.3. Matching AM and welding materials

Integrating additive manufacturing with welding processes necessitates careful consideration of material compatibility. The success of this integration relies on matching the materials used in AM with those suitable for welding. This involves addressing the diverse thermal, chemical, and mechanical properties inherent in both technologies. Metals exhibit varying thermal conductivities and coefficients of thermal expansion. Matching AM and welding metals with similar thermal properties reduces the risk of thermal stresses and distortion during integration. Homogeneity in alloy composition is critical for welding compatibility. AM processes should maintain alloy integrity, ensuring that the material's composition remains consistent throughout the build. Some post-processing steps in AM, such as heat treatments or surface finishes, may affect the material's welding characteristics. Identifying alloys that can withstand such treatments is crucial for maintaining welding integrity. Polymers used in AM should have thermal stability within the temperature ranges associated with welding processes. Thermal degradation or excessive softening during welding may compromise joint strength. Some polymers in AM include additives for specific properties (e.g., reinforcement, flame retardancy). Assessing how these additives may affect the welding process is essential to avoid undesirable outcomes. Matching the melt temperatures of polymers used in AM with those suitable for welding minimizes the risk of uneven melting and improves the overall compatibility of the materials. Conducting weldability tests on AM-produced components with selected alloys or polymers ensures that the materials can withstand the welding process without compromising structural integrity. Analyzing the microstructure of the welded joint can reveal any changes in material properties. This includes assessing grain structures, phase transformations, and potential defects. Testing the mechanical properties of the integrated materials, such as tensile strength and impact resistance, provides insights into the performance of the final product under real-world conditions. By carefully matching AM and welding materials and incorporating thorough testing procedures, manufacturers can ensure a seamless integration that leverages the strengths of both technologies. The subsequent sections will explore real-world applications and delve into the future trends and research directions in the integration of AM with welding processes.

4.4. Optimizing design

AM provides unprecedented freedom in designing intricate and complex geometries. Utilizing this capability enables the creation of structures that were previously challenging or impossible with traditional manufacturing methods. AM allows for the incorporation of internal structures, such as lattices and honeycombs, optimizing components for strength-to-weight ratios. This design freedom can lead to innovative solutions in lightweighting and material efficiency. The ability to tailor designs to specific requirements or individual preferences is a hallmark of AM. This is particularly valuable in industries like healthcare, where personalized medical implants can be created based on patient-specific data. Joint design plays a crucial role in determining the structural integrity of the final product. Optimizing joints ensures that welded components meet the required strength and durability standards. Well-designed joints distribute stresses evenly, reducing the risk of stress concentrations that could lead to premature failure. This is especially critical in applications where components are subjected to dynamic loads or varying environmental conditions. Optimized joint designs take into account the ease of manufacturing, ensuring that welding processes can be carried out efficiently without compromising the quality of the weld. Incorporating gradual transitions in joint designs helps distribute stress more evenly. This is particularly important in welded structures subjected to varying loads. Sharp corners can act as stress concentrators, leading to potential failure points. Rounded or filleted joint designs help reduce stress concentrations and improve overall weld quality. Ensuring proper fit-up of components before welding minimizes gaps and misalignments, contributing to a more robust weld. Proper fit-up is essential for achieving the desired joint strength and integrity [29].

5. CASE STUDIES AND APPLICATIONS

5.1. Aerospace industry

In the aerospace industry, the combination of AM and welding has revolutionized the fabrication of lightweight structures. AM allows for the creation of intricate internal geometries, such as lattice structures, that significantly reduce the weight of components without compromising structural integrity. Welding techniques are then employed to assemble these lightweight components into complex structures, contributing to fuel efficiency and overall performance in aerospace applications. The aerospace industry often requires components tailored to specific aircraft models or mission requirements. AM facilitates the production of customized components with intricate shapes and optimized features. Welding plays a key role in assembling these bespoke components, ensuring a seamless integration that meets the stringent standards of the aerospace sector [30].

5.2. Medical devices

The marriage of AM and welding has had a profound impact on the production of medical implants. Additive Manufacturing allows for the creation of patient-specific implants, precisely tailored to individual anatomies based on medical imaging data. Welding processes are then employed to ensure the structural integrity of these implants, offering a personalized and effective solution for patients in need of orthopedic, dental, or craniofacial implants.

In the medical field, the choice of materials is critical to ensure compatibility with the human body. AM technologies enable the use of biocompatible materials, and welding processes are carefully selected to maintain the integrity of these materials. This integration ensures that medical devices, such as implants and prosthetics, meet both mechanical and biocompatibility requirements [31].

5.3. Automotive sector

The automotive sector benefits from the integration of AM and welding in the creation of hybrid structures. AM allows for the production of lightweight and complex components, while welding techniques are utilized to assemble these components into hybrid structures that combine the strength of traditional materials with the efficiency of lightweight designs. This integration is particularly relevant in the development of electric vehicles and other sustainable automotive solutions.

AM's design freedom facilitates the functional integration of components in the automotive sector. Complex geometries and internal structures can be designed to accommodate multiple functions within a single part. Welding processes are then employed to join these multifunctional components, resulting in a streamlined and efficient automotive manufacturing process [32].

6. FUTURE TRENDS AND RESEARCH DIRECTIONS

6.1. Emerging technologies

Advancements in Material Diversity: As technology evolves, the field of multi-material 3D printing is expected to witness a broader spectrum of materials being integrated into the additive manufacturing process. This includes the simultaneous use of metals, polymers, ceramics, and even electronic components within a single printed object. This trend is set to redefine the possibilities in terms of functionality, durability, and aesthetics.

Improved printing techniques: Ongoing research is focused on refining the techniques involved in multi-material 3D printing. This includes innovations in nozzle design, deposition methods, and software algorithms to precisely control the deposition of different materials. The goal is to enhance the resolution and accuracy of multi-material prints, opening avenues for intricate and diverse applications.

Real-time process optimization: The integration of in-situ monitoring and control aims to provide real-time insights into the additive manufacturing and welding processes. Advanced sensors and monitoring systems will continuously analyze parameters such as temperature, layer adhesion, and structural integrity. Immediate feedback will enable dynamic adjustments during manufacturing, optimizing parameters for improved efficiency and quality.

Machine learning and AI applications: The future of in-situ monitoring involves the incorporation of machine learning and artificial intelligence (AI) algorithms. These technologies will analyze large datasets generated during the manufacturing process, identifying patterns, predicting potential issues, and autonomously making adjustments. This level of intelligence will contribute to the self-optimization of the manufacturing workflow [33].

6.2. Industry 4.0 Integration

Smart factories are set to become highly interconnected ecosystems where machines, devices, and systems communicate seamlessly. The integration of additive manufacturing and welding processes within smart factories will lead to increased automation, allowing for more efficient and flexible production lines.

The use of collaborative robots, or cobots, will play a significant role in smart factories. Cobots can work alongside human operators, assisting in tasks such as material handling, quality control, and even intricate assembly processes. This collaborative approach enhances overall efficiency and safety in the manufacturing environment.

Data-driven decision-making involves leveraging advanced analytics to transform raw manufacturing data into actionable insights. Predictive analytics will be used to foresee potential issues, optimize workflow, and enhance overall process efficiency. This anticipatory approach contributes to reduced downtime and improved resource utilization.

Data-driven decision-making extends beyond the factory floor to the entire supply chain. Integration with suppliers, logistics, and even customer feedback will create a seamless flow of information. This interconnected approach enables adaptive manufacturing strategies, responsive to real-time demand and market fluctuations.

As these emerging technologies and Industry 4.0 principles continue to shape the landscape of additive manufacturing and welding integration, the manufacturing industry is on the cusp of a transformative era characterized by enhanced efficiency, customization, and intelligent decision-making. Researchers and practitioners alike are poised to unlock new potentials and address challenges in this dynamic and evolving field [34].

7. CONCLUSION

7.1. Summary of key findings

The integration of AM with welding processes presents a paradigm shift in manufacturing, offering a host of benefits:

AM provides unparalleled design freedom, allowing for the creation of complex geometries and customized components. The layer-by-layer construction in AM minimizes material waste, while welding streamlines the assembly of components, contributing to overall manufacturing efficiency. The compatibility of AM with a diverse range of materials, including polymers, metals, and composites, coupled with welding, opens avenues for exploring novel material combinations. Case studies in aerospace, medical devices, and the automotive sector demonstrate the successful integration of AM and welding in creating lightweight structures, customized components, patient-specific implants, and hybrid structures. Managing thermal stresses and distortion during the integration of AM with welding remains a challenge. Strategies such as optimized printing parameters, preheating, and post-weld heat treatment are crucial for addressing these issues. Matching materials used in AM with those suitable for welding requires careful consideration. Testing and validation procedures are essential to ensure the structural integrity of integrated components. While AM provides design freedom, optimizing joint designs for welding is critical for achieving structural integrity. Strategies such as gradual transitions, avoiding sharp corners, and proper fit-up contribute to successful joint optimization. Emerging technologies like multi-material 3D printing and in-situ monitoring and control, along with the integration of Industry 4.0 principles, promise to address current challenges and enhance the efficiency and adaptability of the integration process.

7.2. Recommendations for future research

Advanced Thermal Management: Future research should focus on advanced thermal management techniques to further mitigate thermal stresses and distortion during the integration of AM and welding. Comprehensive studies on the compatibility of a wider range of materials used in AM with various welding processes will contribute to a more exhaustive understanding of material interactions. Continued research on joint design optimization, incorporating computational methods and advanced simulations, will further refine the integration process, ensuring robust and efficient joint configurations. Future research should explore new applications in the biomedical field, leveraging the integration of AM and welding for the development of advanced medical devices, implants, and prosthetics. The integration of electronics and sensors into 3D-printed and welded structures opens avenues for the development of smart components with embedded functionalities. Research in this area can lead to innovations in electronics manufacturing. Investigating the environmental impact and sustainability aspects of the integrated AM and welding processes can guide the development of eco-friendly manufacturing practices.

In conclusion, the integration of AM with welding processes is a dynamic and evolving field that has already demonstrated significant benefits across industries.

Addressing current challenges and exploring new applications through continued research and innovation will pave the way for a transformative future in advanced manufacturing.

REFERENCES

[1] Gao W, Zhang Y, Ramanujan D, Ramani K, Chen Y, Williams CB, et al. The status, challenges, and future of additive manufacturing in engineering. *Comput Des* 2015;69:65–89.

[2] Selema A, Ibrahim MN, Sergeant P. Advanced manufacturability of electrical machine architecture through 3D printing technology. *Machines* 2023;11:900.

[3] Mazzoli A. Selective laser sintering in biomedical engineering. *Med Biol Eng Comput* 2013;51:245–56.

[4] Daminabo SC, Goel S, Grammatikos SA, Nezhad HY, Thakur VK. Fused deposition modeling-based additive manufacturing (3D printing): Techniques for polymer material systems. *Mater Today Chem* 2020;16:100248.

[5] Ligon SC, Liska R, Stampfl J, Gurr M, Mülhaupt R. Polymers for 3D printing and customized additive manufacturing. *Chem Rev* 2017;117:10212–90.

[6] Raut LP, Taiwade RV. Wire Arc additive manufacturing: A comprehensive review and research directions. *J Mater Eng Perform* 2021;30:4768–91. https://doi.org/10.1007/s11665-021-05871-5

[7] Fande AW, Taiwade RV. Development of activated tungsten inert gas welding and its current status: A review. *Mater Manuf Process* 2022;37:841–76. https://doi.org/10.1080/10426914.2022.2039695

[8] Roy M, Tran P, Dickens T, Schrand A. Composite reinforcement architectures: A review of field-assisted additive manufacturing for polymers. *J Compos Sci* 2019;4:1.

[9] Williams SW, Martina F, Addison AC, Ding J, Pardal G, Colegrove P. Wire + Arc additive manufacturing. *Mater Sci Technol (United Kingdom)* 2016;32:641–7. https://doi.org/10.1179/1743284715Y.0000000073

[10] Short AB. Gas tungsten arc welding of $\alpha + \beta$ titanium alloys: A review. *Mater Sci Technol* 2009;25:309–24. https://doi.org/10.1179/174328408X389463

[11] Liou FF. Additive manufacturing processes. 2019. https://doi.org/10.1201/9780429029721-6

[12] Wang L, Zhang Y, Hua X, Shen C, Li F, Huang Y, et al. Twin-wire plasma arc additive manufacturing of the Ti–45Al titanium aluminide: Processing, microstructures and mechanical properties. *Intermetallics* 2021;136:107277. https://doi.org/10.1016/j.intermet.2021.107277

[13] Yang D, He C, Zhang G. Forming characteristics of thin-wall steel parts by double electrode GMAW based additive manufacturing. *J Mater Process Technol* 2016;227:153–60. https://doi.org/10.1016/j.jmatprotec.2015.08.021

[14] Ibrahim IA, Mohamat SA, Amir A, Ghalib A. The effect of Gas Metal Arc Welding (GMAW) processes on different welding parameters. *Procedia Eng* 2012;41:1502–6. https://doi.org/10.1016/j.proeng.2012.07.342

[15] Xiong J, Lei Y, Chen H, Zhang G. Fabrication of inclined thin-walled parts in multi-layer single pass GMAW based additive manufacturing with flat position deposition. *J Mater Process Technol* 2017;240:397–403. https://doi.org/10.1016/j.jmatprotec.2016.10.019

[16] Rodrigues TA, Duarte V, Miranda RM, Santos TG, Oliveira JP. Current status and perspectives on wire and arc additive manufacturing (WAAM). *Materials (Basel)* 2019;12. https://doi.org/10.3390/ma12071121

[17] Chen X, Su C, Wang Y, Siddiquee AN, Konovalov S, Jayalakshmi S, et al. Cold Metal Transfer (CMT) based wire and Arc additive manufacture (WAAM) System. *J Surf Investig* 2018;12:1278–84. https://doi.org/10.1134/S102745101901004X
[18] Zhang C, Li Y, Gao M, Zeng X. Wire arc additive manufacturing of Al-6Mg alloy using variable polarity cold metal transfer arc as power source. *Mater Sci Eng A* 2018;711:415–23. https://doi.org/10.1016/j.msea.2017.11.084
[19] Lee SH. Optimization of cold metal transfer-based wire Arc process regression. *Metals (Basel)* 2020.
[20] Shi J, Li F, Chen S, Zhao Y, Tian H. Effect of in-process active cooling on forming quality and efficiency of tandem GMAW–based additive manufacturing. *Int J Adv Manuf Technol* 2019;101:1349–56. https://doi.org/10.1007/s00170-018-2927-4
[21] Liberini M, Astarita A, Campatelli G, Scippa A, Venturini G, Durante M, et al. Selection of optimal process parameters for wire arc additive manufacturing. *Procedia CIRP* 2017;62:470–4. https://doi.org/10.1016/j.procir.2016.06.124
[22] Herzog D, Seyda V, Wycisk E, Emmelmann C. Acta Materialia Additive manufacturing of metals. *Acta Mater* 2016;117:371–92. https://doi.org/10.1016/j.actamat.2016.07.019
[23] Rosli NA, Alkahari MR, Abdollah MF bin, Maidin S, Ramli FR, Herawan SG. Review on effect of heat input for wire arc additive manufacturing process. *J Mater Res Technol* 2021;11:2127–45. https://doi.org/10.1016/j.jmrt.2021.02.002
[24] Patterson T, Hochanadel J, Sutton S, Panton B, Lippold J. A review of high energy density beam processes for welding and additive manufacturing applications. *Weld World* 2021;65:1235–306.
[25] Taiwade RV. Microstructure and mechanical properties of wire arc additively manufactured bimetallic structure of austenitic stainless steel and low carbon steel. *J Mater Eng Perform* 2022;31:8531–41.
[26] Compton BG, Post BK, Duty CE, Love L, Kunc V. Thermal analysis of additive manufacturing of large-scale thermoplastic polymer composites. *Addit Manuf* 2017;17:77–86.
[27] Taiwade RV. Microstructure and mechanical properties of wire Arc additively manufactured bimetallic structure of austenitic stainless steel and low carbon steel. *J Mater Eng Perform* 2022. https://doi.org/10.1007/s11665-022-06856-8
[28] Adil GK, Bhole SD. HAZ hardness and microstructure predictions of Arc welded steels—I. Review of Predictive Models. *Can Metall Q* 1992;31:151–7.
[29] Gibson I, Rosen D, Stucker B, Khorasani M, Gibson I, Rosen D, et al. Design for additive manufacturing. *Addit Manuf Technol* 2021:555–607.
[30] Karayel E, Bozkurt Y. Additive manufacturing method and different welding applications. *J Mater Res Technol* 2020;9:11424–38.
[31] Tilton M, Lewis GS, Manogharan GP. Additive manufacturing of orthopedic implants. *Orthop Biomater Prog Biol Manuf Ind Perspect* 2018:21–55.
[32] Cabigiosu A, Zirpoli F, Camuffo A. Modularity, interfaces definition and the integration of external sources of innovation in the automotive industry. *Res Policy* 2013;42:662–75.
[33] Kanishka K, Acherjee B. Revolutionizing manufacturing: A comprehensive overview of additive manufacturing processes, materials, developments, and challenges. *J Manuf Process* 2023;107:574–619.
[34] Tiwari S. Supply chain integration and Industry 4.0: A systematic literature review. *Benchmarking An Int J* 2021;28:990–1030.

12 Current Scenario, Future Scope, and Challenges in Welding

Nitin Kumar Lautre

1. INTRODUCTION

Welding is an important procedure utilized in many domains, including construction and manufacturing, and the present welding landscape is defined by an increase in demand for high-quality welding services to satisfy the demands of these businesses. With the development of modern welding, robotic welding, and automation, the future of welding looks bright [1]. However, as projects get more complicated, welders confront additional problems, such as the need to understand new technology, comply with safety requirements, and maintain high levels of precision. The welding industry is developing and adapting to modern society's needs, and it is critical for experts in the sector to keep up to speed on the newest innovations in order to face these problems [2]. The salient scope of the discussion of the chapter is presented in Figure 12.1.

2. TYPES OF WELDING TECHNIQUES

New technologies and strategies to improve the welding process are continually appearing in advanced welding techniques. Welding's future promises to be both efficient and accurate as technology advances. Advanced welding equipment and procedures are increasing the quality and productivity of field welding. Laser welding, laser hybrid welding, friction stir welding, electron beam welding, ultrasonic metal welding, and flash butt welding are all examples of sophisticated welding methods. Magnetic, ultrasonic, and solid-state welding methods are being investigated for their potential to produce high-quality welded junctions. Thermal spraying, plasma, and laser surface treatment, surface modification treatments, multilayer structural composites, and nano surface engineering are among the advanced materials and procedures being developed to improve welding and surface technologies [3]. The comparison of the advanced techniques is shown in Table 12.1.

DOI: 10.1201/9781003435884-12

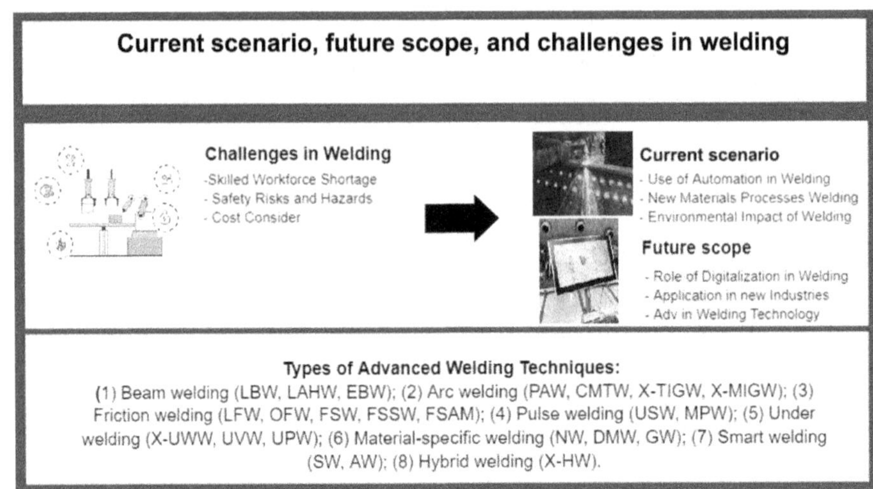

FIGURE 12.1 The challenges and scope of welding

2.1. Beam welding (LBW, LAHW, EBW)

In general, the beam welding process used is determined by the application and the materials being welded. Beam welding is a class of welding procedures that employ a focused beam of energy to connect metal or thermoplastic parts. Beam welding is classified into three types: laser beam welding (LBW), laser-arc hybrid welding (LAHW), and electron beam welding (EBW).

Some of the major traits of each category are shared by both. LBW uses a focused heat source to create narrow, deep welds at rapid speeds. Carbon steels, HSLA steels, stainless steel, aluminum, and titanium may all be welded. Weld quality is good, comparable to electron beam welding. Welding speed is related to the power provided, but it also relies on the kind and thickness of the workpieces. Laser hybrid welding, a subset of LBW, combines the laser beam with another welding technique, such as gas tungsten arc welding (GTAW) or gas metal arc welding (GMAW). Although popular, one of the drawbacks of LBM is that method is not appropriate for welding thick portions. Cracking is a hazard when welding high-carbon steels due to the fast cooling rates. LBW may create more spatter than other welding methods because it is sensitive to joint fit-up and joint gap. The original cost of equipment is significant, as are the maintenance expenses; LBW needs competent operators. The LBW's depth of penetration is determined by the amount of power supplied and the placement of the focal point [4].

The laser beam is combined with another welding method, such as GTAW or GMAW, in LAHW. Attractive for welding together medium-thickness (10–20 mm) plates in a single pass. The advantages of LAHW include fast welding speed, the ability to weld medium-thickness (10–20 mm) plates in a single pass, and the ability to weld in multiple passes. Provides the advantages of both laser and arc welding, resulting in a process that may profit from both. When compared to laser welding, it

TABLE 12.1
Comparison of advanced welding techniques

Weld class	Types of welding	Present	Future	Challenges
Beam welding	LBW, LAHW, EBW.	A high-velocity electron beam connects two materials.	Laser, space, and nuclear applications increase efficiency.	Joints provide solutions to specific weld and sheet metal designs.
Arc welding	PAW, CMTW, X-TIGW, X-MIGW.	Make an arc from numerous sources and unite metals using extreme arc heat.	Mechanical features of welded metal under various welding fillings and currents. High-speed wire feeding and a low gas flow rate.	Control of optical radiation output. Weldment structure and characteristics, as well as intelligent control with physical model development.
Friction welding	LFW, OFW, FSW, FSSW, FSAM.	The heat of rubbing friction of different surfaces in a solid state.	Weld temperature and weld quality are affected by welding parameters. FSW "bobbin stir" variations available.	Welding dissimilar materials.
Pulse welding	USW, MPW.	Ultrasonic or magnetic mode produces pulse mode, frequency, and intermittency.	Enhanced cladding operations.	Ozone and fumes to be reduced.
Under water welding	X-UWW, UVW, UPW.	Controlled cooling rates and power sources are used in wet and local welding.	Welding in a dry spot habitat.	Offshore construction for oil and gas exploration and transportation, ship repair and maintenance, and undersea pipeline repair.
Material-specific welding	NW, DMW, GW.	Investigating innovative welding procedures in accordance with material properties and weld material needs.	Non-traditional welding procedures.	Interactions between material properties.
Smart welding	SW, AW.	Monitoring that is sophisticated and automated in real time.	Robotics and controlled automation. Artificial intelligence and virtual reality are two examples. Future welding will benefit from data analytics.	Initial investment and setup expenses. Scarcity of skills. There are only a few left.
Hybrid welding	X-HW.	A laser and an electrical arc, among other things.	Process enhancement. Weld quality and material compatibility.	The difficulty of the equipment. Distortion and agglomeration.

allows for greater gap tolerances while maintaining high weld speed and penetration. When compared to standard welding processes, the drawbacks of LAHW include greater equipment and maintenance expenses. Skilled operators are required. When compared to laser welding, the procedure has several disadvantages, such as higher sensitivity to joint fit-up and joint gap. Laser welding may create more spatter. The method is ineffective for welding thick parts. [5].

EBW involves bombarding the component to be welded with a highly focused, high-speed beam of electrons, forcing the two components to fuse together as a welded component. Allows for deep penetration, high aspect ratios, and the lowest feasible heat-impacted zone. The welding procedure used is determined by the application and the materials being welded. While LBW is better suited to high-volume applications, EBW is frequently favored for welding high-value components comprised of materials that benefit from a vacuum environment or for welding bigger portions. It is similar to LBW in that a high concentration of beam energy is used at the work surface. It is capable of welding a broad variety of materials, including refractory metals and alloys, and it can even weld dissimilar metals. It is suitable for welding heavy portions as well as welding in a vacuum [6]. Both EBW and LBW are high-energy density welding methods that employ a focused beam of energy to melt the junction of two materials. EBW is often thinner than laser weld. EBW is ideal for low-volume applications. EBW costs more than LBW. If the components are of high value, made of a material that would benefit from the vacuum environment, such as titanium and nickel alloys, the welds are deeper than 1/3 to 1/2 inch, or the laser beam has difficulty coupling with the material being welded, such as aluminum alloys, EBW is often preferred over LBW. Deep penetration and high aspect ratios are feasible with EBW, as is the smallest heat-impacted zone imaginable. Because LBW uses a focused heat source, it is possible to create thin, deep welds while still achieving high welding speeds. Both procedures use a high concentration of beam energy at the work surface, but the way the energy beam is established differs. LBW emits no X-rays and is easily controlled using automation and robots. Both methods are ideally suited to combining components with complicated geometries and are capable of achieving the most demanding standards for the final assembly's metallurgical properties [7].

2.2. Arc welding (PAW, CMTW, X-TIGW, X-MIGW)

Arc welding is a type of welding that uses an electric arc to connect metal pieces. PAW, like TIG welding, generates an electric arc between an anode and a non-consumable electrode inside the torch's body. Weldable materials include stainless steel, copper, nickel, carbon steel, and aluminum. It offers exceptional accuracy and control, making it perfect for welding thin materials [8]. To create the weld, GTAW or TIG) employs a fixed consumable tungsten electrode. Ideal for welding big lengths of stainless steel or nonferrous metals together. Extremely accurate and adaptable, allowing you to combine a variety of tiny and thin materials. Shielded Metal Arc Welding (SMAW) creates welds by using a consumable electrode covered in flux. Carbon steel, alloyed steel, stainless steel, cast iron, and ductile iron are all used to link together. It can also be utilized in nonferrous metal applications such as nickel and copper. MIG welding is similar to flux-cored arc welding (FCAW), except it employs a hollow electrode wire loaded with flux. Welds thick materials and may be utilized outside in windy situations. Other forms of arc welding exist, such as Carbon Arc Welding (CAW) and Submerged Arc Welding (SAW), although the four mentioned above are the most popular. The arc welding process used is determined by the application and the materials being welded [9-11].

2.3. Friction welding (LFW, OFW, FSW, FSSW, FSAM)

Friction welding is a solid-state welding method that creates heat using mechanical friction between workpieces in relative motion. Rotary Friction Welding (RFW) is a welding technique in which one welded element is rotated to the other and forced down. It is utilized in the welding of certain automotive components such as axles, gears, piston rods, and turbine shafts, as well as aluminum copper electrical connections, transition joints that combine incompatible metals, drill bits, and other equipment. Linear Friction Welding (LFW) uses linear motion rather than rotation to create strong connections between two materials by applying heat and pressure. Aerospace applications include integrated bladed disks for jet engines and near-net form blanks for structural elements. Orbital Friction Welding (OFW) is more commonly utilized for plastic joining than metal joining. Although it was not as commonly used, OFW existed for decades before the creation of FSW. Friction Stir Welding (FSW) uses a non-consumable instrument that applies pressure and heat to two pieces of metal before stirring them together to form a connection. It produces high-strength, corrosion- and fatigue-resistant welds, making it excellent for usage in the aerospace, automotive, and electronics manufacturing sectors. FSSW is a version of FSW that produces a "spot" weld by rotating, diving, and withdrawing a non-consumable tool into two workpieces in a lap-joint arrangement. Friction Stir Spot Welding (FSSW) is a kind of FSW that produces a "spot" weld by rotating, plunging, and withdrawing a non-consumable tool into two workpieces in a lap-joint arrangement. Overall, the technique of friction welding used is determined by the individual application and materials being welded [12].

2.4. Pulse welding (USW, MPW)

A range of welding processes known as pulse welding employs a pulsed electric current to connect metal pieces. The pulse welding technique used is determined by the application and the materials being welded. Magnetic Pulse Welding (MPW) is a solid-state technology that employs electromagnetic pressure to accelerate one workpiece and cause it to collide with another. This method produces a metallic bond that is comparable to that produced by explosion welding. The ability to combine incompatible materials is highly recognized. MPW is cost-effective since no filler metals or shielding gases are required, and it reduces environmental expenses. MPW is a cold technique that heats the metal to no more than 30 degrees Celsius, resulting in no heat-affected zone (HAZ) and no metal degradation. MPW may work on any conductive material, including aluminum, copper, magnesium, and steel. MPW has successfully welded aluminum to steel, as well as other dissimilar and equivalent metals. Ultrasonic Welding (USW) generates frictional heat at the joint surface of two workpieces by using high-frequency mechanical vibrations. It is perfect for welding thin materials like plastics and nonferrous metals. Because USW is a low-temperature method, it is appropriate for welding temperature-sensitive materials. USW can be used to fuse different materials together [13-14].

2.5. Under welding (X-UWW, UVW, UPW)

Under welding refers to a collection of welding procedures that use pressure to connect metal pieces without melting the underlying material. The under-welding technique used is determined by the application and the materials being welded. Linear underwater welding (X-UWW) is a specialized welding procedure in which workpieces are pushed together and welded underwater. Ultrasonic Vibration Welding (UVW) generates frictional heat at the joint surface of two workpieces by using high-frequency mechanical vibrations. It is perfect for welding thin materials like plastics and nonferrous metals. Because UVW is a low-temperature method, it is appropriate for welding temperature-sensitive materials. UVW may be used to fuse incompatible materials together. Ultrasonic Peening Welding (UPW) employs ultrasonic energy to generate a high-frequency impact force that peens the workpiece's surface, resulting in a solid-state weld [15].

2.6. Material-specific welding (NW, DMW, GW)

The material-specific welding process used is determined by the application and the materials being welded. Dissimilar Metal Welding (DMW) is a technique for joining two dissimilar structural materials together, such as the junction between a ferritic steel pressure tank and an austenitic stainless steel pipe. DMW can be difficult due to the varying physical, chemical, and mechanical characteristics of the welded components. Before deciding on the optimum approach to link the metals, the filler material and both metals must be analyzed. Green welding is a range of ecologically friendly and sustainable welding processes. These strategies try to lessen the environmental effect of welding by lowering energy usage, eliminating waste, and employing environmentally friendly materials. Nano-welding is a relatively recent welding process that joins nanoscale materials using a concentrated electron beam. The method is still being tested and is being developed for usage in the electronics and medical sectors [16].

2.7. Smart welding (SW)

Smart welding is the application of sophisticated technologies to improve the welding process, such as artificial intelligence (AI), machine learning (ML), and the Internet of Things (IoT). Welding may be made more efficient, precise, and safe by utilizing these technologies. Smart welding may also assist in decreasing waste and increasing weld quality. Civan Lasers and Smart Move GmbH have joined forces to offer novel welding and additive manufacturing solutions to clients. Additive Manufacturing, often known as 3D printing, is the method of manufacturing three-dimensional items by layering material on top of the material. Welding-based additive manufacturing is a proposed technology for creating near-net-shaped metal components using welding procedures. Wire Arc Additive Manufacturing (WAAM) is a welding procedure that produces high deposition rates without regard for quality or size, making it a viable alternative to traditional additive manufacturing processes, particularly for medium and large-scale component production. WAAM is gaining popularity due to advancements in robotic controls and advanced electronic control circuit systems for various welding equipment. The application of modern technologies such as AI, ML, and IoT to optimize the additive manufacturing process is referred to as smart additive manufacturing. These technologies can make additive manufacturing more efficient, precise, and safe. Smart additive manufacturing may also assist in decreasing waste and increasing final product quality [17].

2.8. Hybrid welding (X-HW)

A hybrid welding method combines two or more welding procedures to provide a more accurate and reliable welding process. The benefits of different welding techniques are combined in hybrid welding, which is a promising welding technology. As technology advances, hybrid welding may become more prevalent and increase in precision and efficiency. Currently, laser hybrid welding combines the benefits of laser welding and gas metal arc welding into a single procedure. Hybrid welding combines two or more welding methods, such as laser welding and GTAW or MIG welding, to maximize the benefits of both procedures while minimizing the downsides. Each year, hybrid welding has progressed, becoming a more standardized and dependable welding method. During the forecast period, between 2023 and 2028, the worldwide hybrid welding equipment market is expected to grow at a significant rate. New welding technologies, such as hybrid welding, additive manufacturing, and other ways, are expanding welding's capabilities. The usage of hybrid welding in additive manufacturing is becoming increasingly popular, with the addition of a laser hybrid welding system that combines laser and wire arc welding methods. With developments in automation, robots, and the utilization of data and analytics, hybrid welding may continue to expand and improve [18].

3. CURRENT SCENARIO OF WELDING

The contemporary welding landscape is one of increasing need for qualified welders in areas such as construction, manufacturing, and aerospace. With technological improvements like robots and automation, welders' roles are growing to involve programming and debugging these systems. Furthermore, as safety and regulations become stricter, more emphasis is placed on training and certification programs.

However, the welding business has a number of obstacles, including skilled labor scarcity and the need to keep up with fast-changing technology. In today's atmosphere, professionals in the industry confront both possibilities and obstacles.

3.1. Use of automation in welding

Welding automation is gaining popularity owing to the benefits it provides, such as assuring constant quality, boosting productivity, and minimizing production. Robots have found their way into welding applications as technology has advanced. Robots can do difficult welding jobs with precision, and speed, and without tiredness or breaks. Welding reduces the requirement for specialized labor and manual intervention, boosting workplace safety. It also enables the employment of modern welding techniques like laser welding, which has a greater and faster speed. Despite this, there are obstacles to implementing welding automation, such as the initial investment required for qualified experts to maintain and program the robots, as well as the lack of flexibility [19].

3.1.1. Advantages and disadvantages of automation

The welding industry has seen considerable changes as a result of automation. Automation provides benefits such as enhanced efficiency, less labor, and improved control. Automated welding equipment can work indefinitely without tiring or making mistakes. Furthermore, the workforce may be trained to execute increasingly complicated and cognitively demanding activities, hence enhancing overall performance. However, there are certain drawbacks to welding automation. Machines may be costly to acquire and maintain since they require experienced specialists to operate and repair them. Furthermore, some activities may still need a human touch and skill set, rendering automation ineffective for procedures.

3.2. New materials processes in welding

Welding with new materials processes is an emerging trend in the fabrication business. In response to demands for more efficient and sustainable production, a variety of materials, including improved high-strength steels and lightweight aluminum alloys, have been produced. Advanced welding procedures are required for these materials to achieve high-quality connections as well as increased strength and longevity. Due to their susceptibility to cracking and other flaws, traditional welding procedures have been shown to be inappropriate for welding these materials. Welding technologies such as laser beam (LBW), EBW, resistance welding, and friction are used in new material processes. Manufacturers may enhance weld quality and structural integrity, as well as the productivity of their fabrication process, by adopting these strategies into their welding procedures. Although popular, LBW has restrictions in terms of material thickness that may be welded. Deep welds are limited by LBW, and it is not appropriate for welding thick portions. The thickness of single-pass laser welding is restricted by laser power and welding technology limitations, and the laser penetration level is lower than that of TIG welding. Cracking is a hazard when welding high-carbon steels due to the fast cooling rates. The weld quality is good, comparable to electron beam welding, however, the aspect ratios of LBW welds are extreme, and the number of welding passes depends on the welding speed and the attainable melting. The technique is sensitive to joint fit and gap, and the LBW fit must be almost "perfect". Other welding methods may create more spatter than LBW [20].

3.2.1. Advancements in welding processes

Since its start in the nineteenth century, welding has gone a long way. New welding techniques have been created as a result of technological advancements and the requirement to generate quality welds at a quicker rate. GTAW, GMAW, FCAW, and SAW are examples of these procedures. These innovations not only improved weld quality, but also enhanced production, decreased mistakes, and cut expenses. The advancement of robotic welding technology has increased welding efficiency and productivity. Welding's future is bright, with continuous developments in welding equipment, materials, and processes expected to continue. However, as the demand for high-quality welding goods grows, difficulties like as safety and environmental concerns will need to be addressed.

3.2.2. Regulated metal deposition (RMD)

RMD welding is a high-tech gas metal arc welding method that debuted in 2004. RMD welding is a sophisticated gas metal arc welding technology that produces homogeneous droplet deposition and is employed in a wide range of applications. The approach has been investigated for its parametric effect and optimization of welding performance. To establish a constant metal transfer, the RMD approach predicts and manages the short circuit phase and selectively decreases the welding current. RMD welding is an improvement over regular GMAW. The RMD method delivers homogeneous droplet deposition, allowing the welder to better manage the puddle. RMD welding is utilized in a wide range of applications, including pipe manufacturing, oil and gas pipelines, and steel welding. The RMD approach has been investigated for its parametric effect and optimization of welding performance [21].

3.2.3. Spray transfer

Spray transfer welding is a type of GMAW procedure that includes the transfer of molten metal droplets from the electrode wire to the weld pool. Spray transfer welding is a form of gas metal arc welding technology that is used for fusing heavier materials and delivers good fusion and penetration. It is a more complex welding process than short circuit welding and takes more expertise and experience to conduct. Spray transfer welding delivers a spray of tiny droplets (smaller than the wire diameter) over the arc to the weld pool, which accelerates deposition rates, gives good fusion and penetration, and produces less spatter. Spray transfer welding is used to join heavier materials, generally 1/8 inch or greater in thickness. Spray transfer welding is a method of welding in which the welding wire does not short with the material (it does not come into contact with it). The plasma arc melts the wire, resulting in a shower of small droplets that transfer to the weld pool. Because it is hotter and faster than short circuit welding, spray transfer welding is utilized for particularly heavy steel and production welding. Spray transfer welding is a more complex welding process that takes more expertise and experience to perform than short circuit welding [22].

3.2.4. Surface tension transfer

Lincoln Electric invented surface tension transfer (STT) as a waveform control method for welding gauge material. STT welding is a Lincoln Electric sophisticated waveform control technique that delivers superior weld penetration, lowers spatter and fumes, and is suited for welding on thin materials and joints with open roots or gaps. STT is intended for welding on open root joints, gaps, or thin material with no burn-through. STT is a sophisticated electronic technology that optimizes weld current (waveform) and arc characteristics for a specific job. The surface tension of the filler droplet as it adheres to the weld pool is monitored and regulated using STT technology. STT welding delivers strong weld penetration while requiring less heat input. STT welding lowers spatter and emissions while controlling the current to produce the best results. STT welding is appropriate for welding on thin materials like gauge material, as well as welding on joints with open roots or gaps [23].

3.2.5. Nanometal matrix composite (MMC) powders

Welding uses nanometal matrix composite (MMC) powders to make high-performance materials with increased mechanical and thermal characteristics. The use of nanometal matrix composite powders in welding is an exciting field of study with the potential to develop high-performance materials with better mechanical and thermal characteristics. MMC powders are suitable for use in a wide range of welding processes, including gas tungsten arc welding, laser beam welding, electron beam welding, friction welding, and additive manufacturing. MMCs are metal matrix composite materials made up of ultrafine ceramic or beryllium particles. Powder metallurgy methods are used to create MMC powders, which enable customized mixes of a metal matrix with ultrafine ceramic or beryllium. MMCs may be welded using a variety of processes such as gas tungsten arc welding, laser beam welding, electron beam welding, friction welding, and others. MMCs made from composite powders can be manufactured utilizing additive manufacturing processes like as laser powder bed fusion. Carbon/aluminum MMC junctions of incompatible aluminum alloys may be manufactured using a combination of nanoparticle deposition methods and friction stir spot welding [24].

3.3. Environmental impact of welding

Welding, a necessary process in modern industrial buildings, has a substantial environmental impact that must not be overlooked. Hazardous gases such as carbon dioxide, nitrous oxide, and ozone-depleting chemicals are released throughout the process. Some welding procedures also need the use of hazardous chemical compounds that may be harmful to the environment. Furthermore, welding generates enormous levels of noise pollution as well as a large amount of trash. Welding's environmental effect is expected to rise as industry and building continue to expand globally. As a result, the welding industry must develop and apply sustainable methods to decrease its environmental imprint and create a healthy environment for everybody [25].

3.3.1. Pollution generated by welding

Welding is an industrial procedure that includes melting two or more metals and connecting them together. This process, however, produces enormous pollution in the form of fumes, vapors, and particulate particles. Welding fumes include hazardous elements such as manganese, chromium-nickel, lead, and cadmium, which can cause serious difficulties, skin irritation, and eye damage. Furthermore, when welding in restricted places, the fumes created might displace oxygen, resulting in asphyxiation. Welding also produces a significant quantity of noise pollution, which can lead to hearing damage. As a result, the welding sector must take necessary precautions to reduce the environmental and health risks associated with welding.

3.3.2. Solutions reduce environmental impact

Various methods have been explored to address environmental problems linked with welding. One approach is to use green welding processes, such as friction stir welding, to reduce hazardous emissions. Another solution is to adopt novel welding procedures that do not require the use of shielding gases, which are costly and contribute to ozone layer depletion. Furthermore, several welding consumables have been replaced with environmentally friendly alternatives, reducing the quantity of hazardous waste generated during welding processes. Implementing these solutions improves not just the environment, but also overall productivity, labor costs, and efficiency in the welding sector.

4. FUTURE SCOPE OF WELDING

With new technology developing and breakthroughs being made in established processes, the future of welding appears to be bright. Automation and robots are finding their way into welding, and it is expected that demand for these technologies will expand in the near future. The trend appears to be shifting toward more environmentally friendly welding procedures, with green welding expected to become a significant area of study. Sensors and real-time monitoring are also predicted to increase, resulting in improved quality control and process optimization. Another area where welding will play a key role is additive manufacturing, and D printing has the potential to transform the way we approach welding. Overall, the future scope of seems to be broad and diversified, with plenty of room for development and improvement.

4.1. Role of digitalization in welding

Digitalization has had a huge impact on welding and has altered the sector. Welding has grown more efficient, accurate, and safe as digital technology has advanced. Welding software has enabled virtual testing and optimization of welds, minimizing the requirement for prototypes. Remote welding operations are now feasible thanks to the use of digital communication technologies, allowing activities to be performed in dangerous or difficult-to-reach regions. Additionally, digital sensors and cameras can monitor welding processes for quality control and safety. The incorporation of digital technology in welding has opened up new opportunities for efficiency innovation, but it also faces training and competence hurdles for its successful application [26].

4.1.1. Integration of AI

The incorporation of AI and the IoT has transformed numerous sectors, including welding. Welding equipment may now be configured to monitor and adapt their performance in real time, resulting in higher accuracy and efficiency. Furthermore, IoT may gather data from welding equipment in order to discover patterns and trends, therefore generating insights that can assist enhance the welding process. Welders may operate more effectively, lowering costs and enhancing quality, by combining this technology. Furthermore, the value of AI extends beyond welding, as it can foresee problems and give insights into equipment maintenance and overall process efficiency. While there may be challenges in integrating and deploying these technologies in the sector, the potential advantages are significant.

4.1.2. Implementation of IoT

The use of IoT in advanced welding has the potential to improve welding process efficiency and accuracy. The use of IoT in advanced welding has the potential to increase the process's efficiency, accuracy, and quality. The use of IoT in welding is anticipated to become more prevalent and sophisticated as technology advances. IoT may be utilized to construct welding station monitoring systems that can gather and analyze data in real time to optimize the welding process. The IoT Data-Driven Welding Expert System application can make it easier to gather and integrate welding data. Data-driven welding expert systems may learn and summarize data in order to give insights and recommendations to improve the welding process. Intelligentized Welding Manufacturing and Systems in IoT may be utilized to create multi-agent welding manufacturing systems that increase welding efficiency and accuracy. Capturing and reviewing welding data in IoT provides for the acquisition and evaluation of data generated during welding. This aids in the identification of areas for improvement and the optimization of the welding process. IoT Monitoring and Analyzing Welding Operations has made it simple for fabricators to monitor and evaluate their welding operations by gathering real-time, point-of-weld data that can be utilized to optimize the welding process.

4.2. Application of welding in emerging industries

Welding is a technology that is commonly employed in sectors like construction and automotive. Welding is playing a larger part in modern production as new sectors arise. Welding is important in growing sectors such as renewable energy, 3D printing, and robotics for a variety of applications such as welding for wind turbine components, 3D metal parts, and robotic welding for complicated structures. Welding is not just a connecting method in many cases, but also a tool for invention. As new sectors emerge, welding technologies must adapt and progress to satisfy the growing need for high-quality, efficient, and safe welding procedures [27].

4.2.1. Aerospace, defense, and automotive industry

Welding is used in the aerospace and defense sectors to build airplanes and military vehicles. Welding is vital in the fabrication of high-strength, lightweight materials used in aviation and the military. Welding is utilized in the production of fuselage, wings, landing gear, and components in aerospace. Welding is used in the military sector to produce land, sea, and air-based vehicles, missiles, and munitions. These industries confront problems such as the requirement for sophisticated materials, corrosion resistance, and precise welding. With the development of new materials and joining techniques to meet those difficulties, the future of welding in the aerospace and military sector will continue to change.

4.2.2. Medical industry

The medical business has long been a key consumer of welding technology, particularly in the manufacture of medical equipment and gadgets. Welding is required in the production of surgical tools, implants, and other crucial medical components. In this industry, the necessity for high-quality and dependable welding is crucial, given the lives and well-being of patients who rely on these devices. Welding is also used to create medical equipment, architectural structures, and facilities such as operating rooms, critical care units, and labs. The medical sector is now investing considerably in sophisticated welding technology to increase precision, cleanliness, and efficiency. The area, however, has a number of hurdles, including tight adherence to standards, regulatory regulations, and safety precautions. Welding will likely remain an important technique in the creation of breakthrough medical solutions as the medical sector evolves.

4.3. Advancement in welding technology

The welding business has profited considerably from technological improvements in recent years. Welding operations have gotten more efficient and cost-effective as power sources, control systems, and welding consumables have improved. The introduction of robotic devices is one such technology that has changed welding. These technologies can consistently deliver high-quality results, decreasing the need for human interaction. Furthermore, advances in welding software and simulation tools have improved the precise accuracy of welding processes, allowing welders to plan and manage welding operations more easily. With further developments on the horizon, the sector will continue to profit from technology in the future.

4.3.1. Robotics and automation

Robotics and automation have recently changed the welding business. Welding activities may now be completed more quickly and efficiently with the assistance of robots. These technologies can undertake repetitive or risky operations that would otherwise be performed by humans, therefore improving safety and productivity. Welding robots may be programmed to conduct many procedures with minimum effort, and their speed, accuracy, and consistency make them an excellent choice for welding applications. Automation advancements also made it easier for businesses to maintain production uniformity and high-quality requirements. Despite the benefits, there are some drawbacks to implementing robots and automation in welding, such as the costly initial investment and the requirement for specialized training to run and maintain these systems.

4.3.2. 3D printing in welding

With the growth of technology, the requirement for precision and accuracy in welding has increased, and 3D printing has emerged as a game-changer in this industry. The 3D printing technique entails depositing the metal powder one at a time till the final result is built up. This technique has transformed the manufacture of complicated and intricate patterns with improved quality and minimum waste. Furthermore, 3D printing in welding has opened up unlimited possibilities for new applications, such as the construction of artificial organs and the recycling of metal components. However, there are several issues with D printing in welding, such as scalability and uniformity, that must be addressed before this technique can be extensively used.

5. CHALLENGES IN WELDING

In the current context, the welding business has various challenges, including a scarcity of trained welders, safety issues, and the need for continual training and advancement. Welders must keep up with the latest welding methods and equipment as technology advances, which can be difficult for many. Furthermore, the growing demand for high-quality welded goods and components has put pressure on welding firms to produce on time, resulting in a compromise on safety and quality. Another issue is the increased use of automation and robotics in welding, which has generated worries about the replacement of human welders. To address these issues, welding businesses, training institutes, and governing authorities must work together to safeguard the welding industry's long-term viability and future [28].

5.1. Skilled workforce shortage

The welding sector is now experiencing a skilled manpower shortage, which is affecting firms all around the world. Despite technological advances and breakthroughs, welding is still mostly a manual labor vocation that relies on the knowledge and talents of skilled welders. Unfortunately, there aren't enough trained welders to meet the increasing demand for their skills. This issue is having a substantial impact on welding enterprises, which are struggling to locate and retain the expertise required to meet client expectations and execute orders on time. This scarcity also drives up salaries for experienced welders and adds to the industry's expanding skills gap. If not addressed quickly, this problem might have long-term consequences for the welding industry as a whole.

5.1.1. Encouraging diversity in the welding industry

The welding profession has long been dominated by men, making it difficult for women and other underrepresented groups to enter the field. Diversifying the sector is critical for various reasons. For starters, variety gives new viewpoints and unique ideas that can lead to creativity. Second, it contributes to a more inclusive and inviting industry for all. Third, it can aid in addressing the welding industry's skilled labor crisis. Efforts to encourage diversity include outreach campaigns to schools and organizations, mentorship and training programs for minority groups, and developing a culture of inclusiveness inside welding firms. By fostering diversity in the welding sector, we can secure a bright future for the industry and possibilities for those who want to pursue a career in welding.

5.2. Safety risks and hazards

Welding is a basic technique that is utilized in a wide range of sectors, including building, manufacturing, and fabrication. While it has numerous advantages, it also has some disadvantages. Welders are subjected to high levels of heat, fumes, and radiation, which can cause skin irritation, eye damage, respiratory difficulties, and even cancer. Working with hazardous chemicals and gases, which can be dangerous if not handled properly, is also part of the process. Working with strong tools and equipment can also result in bodily injuries if the proper protective equipment is not worn or the equipment is not utilized appropriately. As a result, it is critical to take the required precautions to safeguard the safety of welders and other workers operating in the proximity.

5.2.1. Measures to promote occupational health

Occupational and safety measures in welding include correct training and equipment. Adequate equipment training decreases the danger of electric shocks. Workers should be educated on how to protect their skin, eyes, and hearing by using protective clothing, goggles, and earplugs. To avoid exposure to fumes, proper measures must be built. Employers should make certain that the workplace is clean, well-lit, and free of hazards. Personal protection equipment (PPE) like as gloves, boots, and helmets should be used on a regular basis to guard against health threats such as burns, wounds, and shock. Appropriate steps to safeguard workers' safety and well-being would boost the efficiency and productivity of the welding sector.

5.2.2. Welding safety

Welding safety is a major problem in the welding business today and in the future. As technology advances, it is critical to ensure that safety regulations stay up with these advancements in order to safeguard employees from injury. Inhaling hazardous vapors and gases can be dangerous. To lessen the risk of developing occupational lung disorders, welders should utilize respiratory protection equipment (RPE), localized exhaust ventilation (LEV), or welding torches equipped with integrated fume extraction. Electric shock is caused by high-voltage electricity, which can produce electric shock if sufficient safety precautions are not taken. Flames and explosions can occur as a result of welding sparks and heat, which can produce flames and explosions if sufficient safety precautions are not taken. Welding produces strong light and heat, which can cause eye and skin damage if sufficient safety precautions are not taken.

If sufficient safety precautions are not taken, the loud levels might cause hearing impairment.

Because of technological advancements, the welding business is always changing and expanding. As manufacturers continue to develop new welding technology, it is critical that safety regulations stay up with these advancements. Welding has become safer and more productive as a result of the use of automation, robots, data, and analytics. It is, however, critical to guarantee that these technologies are utilized securely and that personnel are appropriately taught to use them. As the welding industry transitions to advanced arc welding, automated welding, and cutting-edge metals and alloys, it is critical that safety standards keep up. With increased worries about welders' workplace safety and the hazards of being exposed to toxic welding gases, investments in welding safety are more necessary than ever.

5.2.3. Importance of proper education, training, and equipment

In the welding sector, proper education, training, and equipment are all critical. A solid welding education equips students with the information and abilities they need to complete welding activities safely and successfully. Proper training is required for welding machine operators to perform first-rate welds safely and correctly. Welding quality and productivity in field applications may be improved using advanced welding equipment and methods. New-generation enterprises are enhancing their sustainability and productivity by using these methods, which increase the quality of the working environment and circumstances. A solid welding education is one of the foundations required for success in the welding profession. Welding education equips students with the information and abilities required to complete welding jobs safely and successfully. Welding education may introduce students to welding as a possible career path. Proper training is required for welding machine operators to perform first-rate welds safely and correctly. Welding training may give students hands-on experience as well as simulations of various welding procedures. Welding training may help students acquire welding skills more quickly while also promoting a safe welding workplace. Welding quality and productivity in field applications may be improved using advanced welding equipment and methods. Welding has become safer and more productive as a result of the use of automation, robots, data, and analytics. Welding equipment, such as welding torches with integrated fume extraction, can lower the chances of acquiring occupational lung disorders.

5.3. Cost consideration

In the welding business, is a crucial aspect to consider when selecting welding processes, equipment, and consumables. While certain welding procedures may provide more precision, they may also be more expensive. Additionally, the cost of training and supplying suitable safety equipment and protective clothing for welders can add to the overall costs. However, cost should not be the primary consideration when selecting welding procedures, since reduced costs may result in poorer welds and decreased efficiency. Finding the correct balance between cost and quality is critical for businesses to remain competitive in the welding industry [29].

5.3.1. Impact of welding on project costs

Welding is a critical technique in the construction, production, and repair of items in a variety of sectors. It has a direct influence on project costs since welding is a time-consuming, labor-intensive activity that necessitates expert people and expensive equipment. The welding process may dramatically raise the price of materials, equipment, and labor. Welding is a vital aspect that impacts product quality, and any mistakes or faults result in increased costs to correct such concerns. As a result, it is critical to guarantee that the welding process is efficient and effective in order to decrease project costs and enhance overall profitability.

5.3.2. Cost-effective solutions for welding projects

Welding projects may be expensive, so finding cost-effective solutions is a priority. There are various options for cutting costs without losing quality, ranging from selecting the correct materials to improving welding procedures. Automation and robots have gained in popularity in recent years, providing businesses with extremely efficient and cost-effective options for welding projects. Furthermore, using alternative welding techniques like friction stir welding and laser welding can reduce total project expenses. Finally, outsourcing welding jobs to competent and experienced personnel can provide cost-effective options while delivering excellent outcomes. Finally, knowing your project's requirements and employing the appropriate techniques may result in considerable cost reductions in welding projects.

6. CONCLUSION

Welding is becoming an essential component of contemporary production processes. Welding is currently undergoing a significant move toward automation and robotics. The demand for better productivity, efficiency, and safety in welding processes has prompted this transition. The development of sophisticated materials, such as composites and alloys, necessitates the development of novel welding processes and technologies. These advancements, however, offer substantial hurdles, such as the requirement for experienced employees and the high cost of equipment and upkeep. Another significant issue in welding is environmental sustainability. Welding procedures produce enormous amounts of heat, noise, and fumes, all of which can harm the environment and human health. To solve this issue, scientists are investigating novel welding procedures that decrease emissions and waste. Furthermore, there is an increasing demand for environmentally friendly welding methods such as recycling and reusing welding consumables.

Welding and its surroundings are continually developing, with new technologies and procedures being developed to better the welding process. Welding's future promises to be both efficient and accurate as technology advances. New welding technologies, such as hybrid welding, additive manufacturing, and other ways, are expanding welding's capabilities. Various welding processes collaborate to provide a more accurate and reliable welding process. Recent advances in sophisticated materials and techniques, such as thermal spraying, plasma and laser surface treatment, surface modification treatments, multilayer structural composites, and nano surface engineering, shed light on recent advances in welding and surface technologies. The best metals for welding are determined by the job and the technique utilized. Automation and robotics advancements, artificial intelligence, and virtual reality promise to make the future of welding cutting-edge and extremely efficient. During the forecast period, between 2023 and 2028, the worldwide hybrid welding equipment market is expected to grow at a rapid pace. With developments in automation, robots, and the utilization of data and analytics, hybrid welding may continue to expand and improve. Welding techniques such as laser welding, electron beam welding, plasma welding, ultrasonic welding, diffusion welding, and friction stir welding will be studied further in order to produce satisfying and high-quality welded parts. With developments in robotic controls and contemporary electronic control circuit systems for various welding equipment, the use of additive manufacturing in welding is becoming increasingly frequent.

REFERENCES

[1] Xu, Fengjing, Yanling Xu, Huajun Zhang, and Shanben Chen. 2022. "Application of Sensing Technology in Intelligent Robotic Arc Welding: A Review." *Journal of Manufacturing Processes* 79: 854–80. https://doi.org/10.1016/j.jmapro.2022.05.029

[2] Chakradhar, Ritesh, Jorge Ortega-Moody, Kouroush Jenab, and Saeid Moslehpou. 2022. "Improving the Quality of Welding Training with the Help of Mixed Reality along with the Cost Reduction and Enhancing Safety." *Management Science Letters* 12 (4): 321–30. https://doi.org/10.5267/j.msl.2022.4.002

[3] Salam, R., and A. Adi. 2023. "Learn Welding Techniques to Improve Work Skills in Vocational School Students." *Journal PKM*.

[4] Cho, Dong Min, Jin Sung Park, Seung Gab Hong, and Sung Jin Kim. 2023. "Corrosion Behaviors According to the Welding Process of Super duplex Stainless Steel Welded Tubes: Gas Tungsten Arc Welding vs. Laser Beam Welding." *Corrosion Science* 216 (111108): 111108. https://doi.org/10.1016/j.corsci.2023.111108

[5] Du, Lingzhi, Zhibin Yang, and Xing Wang. 2023. "Welding Characteristics of Laser-MIG Hybrid Welding of Arc-Welded Aluminum Profiles for High-Speed Trains." *Materials* 16 (1): 404. https://doi.org/10.3390/ma16010404

[6] Salehi-Shabestari, Ali, Arash Khakzadshahandashti, and Mohammad Reza Rahimipour. 2022. "Numerical Modelling of Electron Beam Welding (EBW) of Zhs6u Superalloy and Its Experimental Validation." *Materials at High Temperatures* 39 (1): 12–20. https://doi.org/10.1080/09603409.2021.2002237

[7] Zhao, Yanhua, Wenyan Sun, Qian Wang, Yujing Sun, Jiwen Chen, Chuanbin Du, Hongyu Xing, Nan Li, and Wenhao Tian. 2023. "Effect of Beam Energy Density Characteristics on Microstructure and Mechanical Properties of Nickel-Based Alloys Manufactured by Laser Directed Energy Deposition." *Journal of Materials Processing Technology*, no. 118074: 118074. https://doi.org/10.1016/j.jmatprotec.2023.118074

[8] Maksymov, S. Yu, E.O. Paton Electric Welding Institute, NASU, D. M. Krazhanovskyi, Yu A. Shepelyuk, S. V. Osynska, E.O. Paton Electric Welding Institute, NASU, E.O. Paton Electric Welding Institute, NASU, and E.O. Paton Electric Welding Institute, NASU. 2022. "Effect of Parameters of Pulsed-Arc Welding on the Formation of Weld

Metal and Microstructure of Heat-Affected Zone of 09G2S Steel." *Paton Welding Journal* 2022 (3): 19–26. https://doi.org/10.37434/tpwj2022.03.02

[9] Abbas, Mahmoud, Ahmed Sayed Hamdy, and Essam Ahmed. 2020. "The Comparison of Gas Tungsten Arc Welding and Flux Cored Arc Welding Effects on Dual Phase Steel." *Materials Research Express* 7 (3): 036523. https://doi.org/10.1088/2053-1591/ab7f5f

[10] Alipooramirabad, H., N. Cornish, R. Kurji, A. Roccisano, and R. Ghomashchi. 2023. "Quenched and Tempered Steels Welded Structures: Modified Gas Metal Arc Welding-Pulse Vs." *Shielded Metal Arc Welding*. Metals.

[11] Neves, Alisson Caetano, João Roberto Sartori Moreno, Celso Alves Corrêa, and Emillyn Ferreira Trevisani Olívio. 2021. "Study of Arc Welding Stability in Flux Cored Arc Welding Process and Pulsed Continuous Current." *Welding International* 35 (4–6): 158–69. https://doi.org/10.1080/09507116.2021.1971936

[12] Casalino, Giuseppe. 2020. "Recent Achievements in Rotary, Linear and Friction Stir Welding of Metals Alloys." *Metals* 10 (1): 80. https://doi.org/10.3390/met10010080

[13] Li, Jishuai, Rijia Nirina Raoelison, Thaneshan Sapanathan, and Mohamed Rachik. 2023. "Determining the Weldability Window Based on the Interface Morphologies Formed during High-Speed Collision in Magnetic Pulse Welding." *Science and Technology of Welding & Joining* 28 (4): 259–67. https://doi.org/10.1080/13621718.2022.2151733

[14] Zhi, Qian, Yongbing Li, Peng Shu, Xinrong Tan, Caiwang Tan, and Zhongxia Liu. 2022. "Double-Pulse Ultrasonic Welding of Carbon-Fiber-Reinforced Polyamide 66 Composite." *Polymers* 14 (4): 714. https://doi.org/10.3390/polym14040714

[15] Baradarani, F., S. Emami, A. Mostafapour, and Khan F. Md. 2021. "Influence of Ultrasonic Vibration on the Microstructure and Texture Evolution of AZ91 Magnesium Alloy during Ultrasonic-Assisted Friction Stir Welding." *Welding in the World* 65 (12): 2371–82. https://doi.org/10.1007/s40194-021-01157-5

[16] Kim, Tae Won, and Hae Woon Choi. 2021. "Study on Laser Welding of Al-Cu Dissimilar Material by Green Laser and Weld Quality Evaluation by Deep Learning." *Journal of Welding and Joining* 39 (1): 67–73. https://doi.org/10.5781/jwj.2021.39.1.8

[17] Dupriez, Nataliya Deyneka. 2021. "Smart Monitoring of Battery Welding Processes: Sensor Systems for Laser Welding Applications along the Battery Production Chain: From Cell to Package." *PhotonicsViews* 18 (5): 46–50. https://doi.org/10.1002/phvs.202100045

[18] Xiu, Qingpeng, Zuming Liu, and Xingchuan Zhao. 2023. "KTIG–MIG Hybrid Arc Welding Process." *Science and Technology of Welding & Joining*, 1–10. https://doi.org/10.1080/13621718.2023.2213916

[19] Mazur, A. A., E.O. Paton Electric Welding Institute, NASU, O. K. Makovetskaya, S. V. Pustovojt, E.O. Paton Electric Welding Institute, NASU, and E.O. Paton Electric Welding Institute, NASU. 2017. "The Main Tendencies of Development of Automation and Robotization in Welding Engineering (Review)." *Paton Welding Journal* 2017 (6): 2–8. https://doi.org/10.15407/tpwj2017.06.01

[20] Fujii, Hidetoshi. 2020. "New Joining Methods for (Iron-Based) Automotive Materials: ISMA-Results." *Journal of the Japan Welding Society* 89 (6): 425–29. https://doi.org/10.2207/jjws.89.425

[21] Das, Subhash, Jay J. Vora, Vivek Patel, Wenya Li, Joel Andersson, Danil Yu Pimenov, Khaled Giasin, and Szymon Wojciechowski. 2021. "Experimental Investigation on Welding of 2.25 Cr-1.0 Mo Steel with Regulated Metal Deposition and GMAW Technique Incorporating Metal-Cored Wires." *Journal of Materials Research and Technology* 15: 1007–16. https://doi.org/10.1016/j.jmrt.2021.08.081

[22] Xiao, Lei, Ding Fan, Jiankang Huang, Shinichi Tashiro, and Manabu Tanaka. 2022. "Mild Steel Metal Rotating Spray Transfer Behavior in Magnetically Controlled Gas Metal Arc Welding." *Materials Today. Communications* 31 (103352): 103352. https://doi.org/10.1016/j.mtcomm.2022.103352

[23] Liu, Wang, Zhijiang Wang, Zhendong Chen, Huawei Liu, Shaojie Wu, Dongpo Wang, and Shengsun Hu. 2022. "Sensing and Characterization of Backside Weld Geometry in Surface Tension Transfer Welding of X65 Pipeline." *Journal of Manufacturing Processes* 78: 120–30. https://doi.org/10.1016/j.jmapro.2022.04.011

[24] Moreira, Roberta Cristina Silva, Oksana Kovalenko, Daniel Souza, and Ruham Pablo Reis. 2019. "Metal Matrix Composite Material Reinforced with Metal Wire and Produced with Gas Metal Arc Welding." *Journal of Composite Materials* 53 (28–30): 4411–26. https://doi.org/10.1177/0021998319857920

[25] Knysh, V. V., E.O. Paton Electric Welding Institute, NASU, S. O. Solovei, L. L. Nyrkova, A. O. Gryshanov, V. P. Kuzmenko, E.O. Paton Electric Welding Institute, NASU, E.O. Paton Electric Welding Institute, NASU, E.O. Paton Electric Welding Institute, NASU, and E.O. Paton Electric Welding Institute, NASU. 2020. "Impact of High-Frequency Peening and Moderate Climate Atmosphere on Cyclic Fatigue Life of Tee Welded Joints with Surface Fatigue Cracks." *Paton Welding Journal* 2020 (1): 37–42. https://doi.org/10.37434/tpwj2020.01.05

[26] Tsuzuki, Ryoichi. 2022. "Development of Automation and Artificial Intelligence Technology for Welding and Inspection Process in Aircraft Industry." *Welding in the World* 66 (1): 105–16. https://doi.org/10.1007/s40194-021-01210-3

[27] Ahmed, M. M., M. M. El-Sayed Seleman, D. Fydrych, and G. Çam, 2023. Friction Stir Welding of Aluminum in the Aerospace Industry: The Current Progress and State-of-the-Art Review. *Materials*, 16(8): 2971.

[28] Francis, J. A. 2022. "A Perspective on Welding Technology Challenges in the Nuclear Sector." *Science and Technology of Welding & Joining* 27 (4): 309–17. https://doi.org/10.1080/13621718.2022.2047407

[29] Joshuva. 2020. "Wastage Cost Analysis of Fillet Welding on Boiler." *Journal of Advanced Research in Dynamical and Control Systems* 12 (01-Special): 505–11. https://doi.org/10.5373/jardcs/v12sp1/20201097

ABBREVIATIONS

AM Additive manufacturing
AR Augmented reality
AW Additive welding
AUW Automated welding
CMTW Cold metal transfer welding
DMW Dissimilar metal welding
EBW Electron beam welding
FSW Friction stir welding
FSAM Friction stir additive manufacturing
FSSW Friction stir spot welding
FW Friction welding
GMAW Gas metal arc welding
GW Green welding
HW Hybrid welding

LAHW	Laser-arc hybrid welding
LBW	Laser beam welding
LFW	Linear friction welding
LPBF	Laser powder bed fusion
MMC	Metal matrix composites
MPW	Magnetic pulse welding
NW	Nano-welding
OFW	Orbital friction welding
PAW	Plasma arc welding
RMD	Regulated metal deposition
SW	Smart welding
USW	Ultrasonic welding
UMFW	Under magnetic field
UPW	Under powder welding
UVW	Under vacuum welding
VR	Virtual reality
X-HW	Various—Hybrid welding
X-MIGW	Various—Metal Inert gas welding
X-TIGW	Various—Tungsten Inert Gas welding
X -UWW	Various—Underwater welding

Index

Note: Page numbers in *italics* and **bold** denote figures and tables, respectively.

A

AA2219-T8 alloy 143–145, *144*
AA5059 alloy 141
AA6000 series 147
AA6005-T5 aluminum alloy 184, 189
AA7055 alloy 143, 145
acoustic emission testing **81**, 82–83
activated-tungsten inert gas (A-TIG) welding 1–2
additive manufacturing (AM) technology 154, 162, 175, 198, 208–209
  advanced welding techniques 203
  case studies and applications 205–206
    aerospace industry 205–206
    automotive sector 206
    medical devices 206
  challenges and solutions 203–205
    heat-affected zones and distortion control 204
    matching AM and welding materials 204–205
    optimizing design 205
    thermal stresses management 203–204
  cold metal transfer (CMT) 202
  future trends and research directions 206–207
    emerging technologies 206–207
    Industry 4.0 integration 207
  GMAW-based WAAM 202
  GTAW-based WAAM 201
  and laser welding 175
  materials 200–201
  objectives of 198–199
  overview of 199–201
  PAW-based WAAM 201–202
  tandem GMAW 202
  techniques 200
  welding parameters 202–204
  welding processes 201–203
aerospace industry, application of welding in 13, 223
  and additive manufacturing integration 205–206
  use of laser beam welding in 174
AI-based welding techniques 2, 207, 216, 222
air hole, in laser welding 27
Al7050-T76 159
aluminum 139–140
  metal inert gas (MIG) welding for 145–148
  tungsten inert gas (TIG) welding for 141–145
aluminum 6xxx grade extrusions 14
aluminum alloys 14
  and solidification cracking 104–105, **105**
  2219 Al-Cu alloy 141
arc strike 69–73, *72*, **72**
arc welding 14, 38–39, **213**, 214–215
  arc initiation techniques 39–40
    electron beam welding 43
    field start method 42, *42*
    pilot arc 42
    plasma arc torch 43
    scratch start method 42
    touch start method 40–41, *41*
  arc temperature 48–50
  of dissimilar metallic materials 173–174
  electric arc generation 38
  electrode contribution 39
  emission of free electrons 43
    ionization potential 43
    mechanisms responsible for 43–44
    work function 43
  excitation and ionization 39
  heat distribution and transfer 39
  heat generation 38–39
  heat transfer 50–51, *51*
    arc heating 51–52
    heat conduction equation 53
    heat transfer by convection 53
    radiative heat transfer 53
    resistance heating 52–53
    through pulsed gas metal arc welding (P-GMAW) 54
  low ionization potential elements, method involving 45
  low power factor, approach utilizing 45–46, *46*
  maintenance of arc 44
    arc control technology 45
    arc length control 44
    electrode and torch cooling 45
    electrode consumption 44
    electrode position and angle 44
    operator skill 45
    shielding gas and flux 45
    welding current and voltage control 44
  melting and weld pool formation 39
  metal transfer processes 53–54
    globular transfer 53
    hot start and cold start transfers 54
    pulsed transfer 54

short-circuiting transfer 54
spray transfer 53
plasma arc welding (PAW) 13, 201–202, 214
plasma formation 38
shielding gas 39
welding arc characteristics 46, *47*, *48*
drooping characteristic zone 47
flat characteristic zone 47
rising characteristic zone 47–48
argon 181
augmented reality (AR), for training and simulation 31
austenitic-austenitic stainless steel 127–128, *128*
austenitic-duplex stainless steel 128–129, *129*
austenitic-ferritic stainless steel 129–130, *130*
austenitic stainless steels (ASSs) 95, 102–103, 106, 118–120, *119*
autogenous welding 100
automated sensing, in robotic welding 31
automation of welding processes 148–149, 218, 223–224
automotive industry, application of welding in 13–14, 92–93, 223
and additive manufacturing integration 206
use of hybrid laser-arc welding in 191
use of laser beam welding in 174

B

beam welding 212–214, **213**
see also electron beam welding (EBW); laser-arc hybrid welding (LAHW); laser beam welding (LBW)
bimetallic corrosion *see* galvanic corrosion, effect of filler material on
burn-through defects *72*, **73**, 74

C

CAM control, in laser welding 25
carbon and low alloy steel, and solidification cracking 105–106
carbon arc welding 214
carbon manganese steels 98–99
carbon migration 91
challenges in welding *212*, 224–227
cost consideration 226–227
cost-effective solutions 227
impact of welding on project costs 227
encouraging diversity in the welding industry 224–225
safety risks and hazards 225–236
measures to promote occupational health 225
proper education, training, and equipment, importance of 226
welding safety 225–226
skilled workforce shortage 224
chromium 95
cleaning inadequacy, defects due to 63
coarse-grain HAZ (CGHAZ) 121, 125–126, 130, *130*, 189
cold cracks *75*, **76**, 77–78
cold metal transfer (CMT) welding
in additive manufacturing 202
of ferritic stainless steels (FSSs) 121
collaborative robots (cobots) 207
collapse, in laser welding 27
computational fluid dynamics (CFD) 160
conduction laser welding 168–169, *169*, **170**
construction industry, use of welding in 13
continuous laser 25
copper 92
copper nickel alloys 159
corrosion 91, 93–98, *97*
cost consideration 226–227
cost-effective solutions 227
project costs 227
couple-thermo flow model 160–161
cracks 74–76, *75*
case study 102–103
cold cracks **76**, 77–78
crater cracks **75**, *75*, 77
ductility-dip cracking 96–97
hot cracks **76**, 77
in laser welding 27
solidification *see* solidification cracking
Cr-Mn-Ni-N ASS 127
Cr-Ni ASSs 127
current scenario of welding *212*, 217
automation in welding 218
advantages and disadvantages of 218
environmental impact 220–221
pollution 221
solutions reduce 221
new materials processes 218–220
advancements 219
nanometal matrix composite (MMC) powders 220
regulated metal deposition 219
spray transfer 219
surface tension transfer 220

D

data-driven decision-making 207
defects
arc strike 69–73, *72*, **72**
burn-through *72*, **73**, 74
and cleaning inadequacy 63
cracks *see* cracks
detection methods 78–87

acoustic emission testing **81**, 82–83
dye penetrant test 79, **79**, *80*
eddy current testing **85**, 85–87, *86*
infrared thermography **86**, *86*, 87
magnetic particle testing 80–82, **81**
radiography testing **82**, *83*, 83–84
ultrasonic testing **84**, 84–85
visual inspection **78**, 78–79
distortion **73**, 74, *75*
excess penetration 68, *69*, **70**
excess reinforcement 68, **68**, *69*
improper electrode selection **77**, 78
and improper joint preparation 64
and inadequate material preparation 64
incomplete fusion 66–68, **67**, *69*
incomplete penetration 68, *69*, **70**
and incorrect parameters 63
misalignment 74, **74**, *75*
overlap 65, *65*, **66**
porosity **64**, 64–65, *65*
slag inclusion 68–69, **71**, *72*
spatter 69, **71**, *72*
undercut 65, *65*, **66**
underfill *65*, 66, **67**
defense sector, application of welding in 223
diffusion welding 13, 14
digitalization in welding 221–222
implementation of IoT 222
integration of AI 222
dilution with filler metal 103–104, *104*
direct metal laser sintering (DMLS) 200
dissimilar materials joining
aluminum alloy joints 146
using FSW 159–160
using laser beam welding 171
dissimilar metal welding (DMW) 115, 216
distortion defects **73**, 74, *75*
ductility-dip cracking 96–97
duplex-martensitic stainless steel 130–131
duplex stainless steel (DSS) 101–102, 123–125, *124*
dye penetrant test 79, **79**, *80*

E

eddy current testing **85**, 85–87, *86*
electric arc generation 38
electrode contribution, in arc welding process 39
electrode selection, and defects **77**, 78
electron beam melting (EBM) 200
electron beam welding (EBW) 10, *10*, 13, 27, 115, 212, 214
and additive manufacturing 203
advantages and disadvantages of 28
applications 28–29
in arc welding 43
and laser beam welding (LBW), comparison between **23**
working principle 27–28, *28*
emerging industry, use of welding in 222–223
aerospace, defense, and automotive industry 223
hybrid welding 192
laser beam welding 174
medical industry 223
environmental impact 220–221
pollution 221
solutions reduce 221
excess reinforcement defects 68, **68**, *69*
excitation, in arc welding process 39
explosive welding *8*, 8–9, 33
applications 35
disadvantages of 35
setup parts 33–34
working principle 34, *34*

F

ferrite number (FN) 99, 108–109, **111**
ferritic stainless steel (FSS)
joining to tantalum 92
structure-property correlation 120–123, *122*
fiber lasers 26, 170, 172, 174
field emission 43
field start method, in arc welding 42, *42*
filler materials 90–91
dilution and its effect 103–104, *104*
fusion region, effect on 99–101
case studies 101–103
galvanic corrosion, effect on 93–98, *97*
selection of filler metal 91–93
corrosion properties 93
metallurgical compatibility 92
physical properties 92–93
and solidification cracking 98–99, **100**, 104
aluminum alloy 104–105, **105**
carbon and low alloy steel 105–106
weld metal constitution, prediction of 106–112
case study 109–112
fine-grained HAZ (FGHAZ) 121, 125, *126*, 130, 189
finite element analysis (FEA) 160, *161*
fixtures elimination, using robotic welding 31
flux-assisted welding 4
flux bounded TIG (FB-TIG) 2
flux-cored arc welding (FCAW) 215
flux zone TIG (FZ-TIG) 2
free electrons emission, in arc welding 43–44
field emission 43
ionization potential 43

secondary emission 44
thermo-ionic emission 43
work function 43
friction stir spot welding (FSSW) 156, 215
friction stir welding (FSW) 2, 11, 115, 153–154, 162–163, 215
and additive manufacturing 203
advantages and disadvantages **24**
for aluminum 141
applications 30
ASS welded using 120
challenges and approaches *156*
design and parameters 154–159, *155*
DSS welded using 124–125
energy-assisted 2
FSS welded using 122
numerical techniques in 160–162
possible future work avenues 162
processing parameters of 30
similar and dissimilar metal joining 159–160
working process 29, *29*
friction welding (FW) 2, 11–12, *12*, **213**, 215
fused deposition modeling (FDM) 200
fusion region, filler metal effect on 99–100
autogenous welding 100
case studies 101–103
constitution diagrams 107–109
heterogenous welding 101
homogenous welding 100
nickel-chromium diagram 107
WRC-1988 diagram 108–109
WRC-1992 diagram 109, 111
future scope of welding 14–15, *212*, 221–224
advancement in technology 223–224
robotics and automation 223–224
3D printing in welding 224
digitalization in welding 221–222
implementation of IoT 222
integration of AI 222
emerging industries, application in 222–223
aerospace, defense, and automotive industry 223
medical industry 223

G

galvanic corrosion, effect of filler material on 93–98, *97*
gap-bridging 187–188
gas laser 26
gas metal arc welding (GMAW) (metal inert gas (MIG) welding) 5–6, *6*, 138–140, *140*, 212, 215
for aluminum 145–148
ASS welded using 120
-based WAAM 202
FSS welded using 121–122
and laser, hybrid welding 32
robotic advancement in *148*, 148–149
UFC-MIG technique 146, *146*
gas tungsten arc welding (GTAW) (tungsten inert gas (TIG) welding) 1–2, 4–5, *5*, 115, 136–138, *140*, 141, 212
for aluminum 141–145
ASS-FSS joints welded using 129
ASS welded using 119–120
-based WAAM 201
DSS welded using 125
FSS welded using 121
and laser beam, hybrid welding 32
MSS welded using 125
robotic advancement in *148*, 148–149
globular transfer, in arc welding 53
green welding (GW) 216

H

heat input (HI) equation 116
heat source, welding process based on 3
electron beam welding 10, *10*
explosive welding *8*, 8–9
gas metal arc welding 5–6, *6*
gas tungsten arc welding 4–5, *5*
laser beam welding 9–10
oxy-acetylene gas welding 7–8
resistance welding 6–7, *7*
shielded metal arc welding 3–4, *4*
thermite welding 9
heat transfer, in arc welding 50–53
arc heating 51–52
heat conduction equation 53
heat transfer by convection 53
radiative heat transfer 53
resistance heating 52–53
through P-GMAW *see* pulsed gas metal arc welding (P-GMAW)
helium 136
heterogenous welding 101
high-energy density beam welding 2
high-velocity impact welding 2
homogenous welding 100
hot cracks *75*, **76**, 77
hot start and cold start transfers, in arc welding 54
hybrid welding 31–32, **213**, 217
areas of applications 191–192
disadvantages of 32
laser-arc welding 181–182
angle of electrode 187
energy input 183–184
focal point position 186–187

joint gap 187–188
laser beam and arc, interaction between 182–183
microstructure-mechanical property relationship 189–190
relative position of the laser and the arc torch 185–186
root hump formation 188
welding speed 184–185
laser-friction stir welding 190–191
laser-MAG welding 192
laser-MIG/MAG hybrid welding 32
laser-plasma hybrid welding 32
laser-TIG hybrid welding 32
MSS welded using 125

I

incomplete fusion defects 66–68, **67**, *69*
Inconel alloys 102–103
infrared thermography **86**, *86*, 87
inter-critical HAZ (ICHAZ) 125, *126*, 189
intermetallic compounds (IMCs) 173
Internet of Things (IoT) 31, 216, 222
ionization, in arc welding process 39, 43, 45

K

keyhole laser welding *169*, 169–170, **170**

L

laser-arc hybrid welding (LAHW) 181–182, 212, 214
angle of electrode 187
energy input 183–184
focal point position 186–187
joint gap 187–188, *188*
laser beam and arc, interaction between 182–183
microstructure-mechanical property relationship 189–190
relative position of the laser and the arc torch 185–186
root hump formation 188
welding speed 184–185
laser beam welding (LBW) 9–10, 13, 22–23, *24*, 115, 167, *168*, 177, 212, 218–219
and additive manufacturing 203
advantages and disadvantages of 26
air hole due to 27
applications of 27, 174–175, *176*
challenges in 171
common metallurgical defects 171–172
high-cost concern 173
joining of dissimilar materials 171
laser sensitivity to materials *172*, 173
residual stress development 172–173
safety concern 173
collapse due 27
conduction mode of operation 25, 168–169, *169*, **170**
continuous mode of operation 170
crack due to 27
defects and remedies 27
dissimilar welding 173–174
and EBW, comparison between **23**
keyhole mode of operation 25, *169*, 169–170, **170**
parameters 170–171, *171*
pulsed mode 170
rapid heating and cooling technique 172
splash due to 27
types of laser 25–26
continuous laser 25
fiber laser 26
gas laser 26
pulsed laser 25
solid-state laser 26
undercut due to 27
welding steel with other alloys 173–174
working principle 23–25
laser-friction stir hybrid welding 190–191
laser-MIG/MAG hybrid welding 32
laser-plasma hybrid welding 32
laser technology 2
laser-TIG hybrid welding 32
linear friction welding 12
linear friction welding (LFW) 215
linear underwater welding (X-UWW) 216
liquation cracking, effect of filler material on 98–99, **100**, 105–106
low-Ni ASS 127

M

machine learning 207, 216
magnetic particle testing 80–82, **81**
magnetic pulse welding (MPW) 215–216
manual metal arc welding (MMAW) 3–4, *4*
martensitic stainless steel 125–127, *126*
material diversity, advancements in 206–207
material-specific welding **213**, 216
medical sector, application of welding in 223
and additive manufacturing integration 206
use of laser beam welding in 174–175
melting and weld pool formation, in arc welding process 39
melting energy increment 183
metal additive manufacturing 200

metal inert gas (MIG) welding *see* gas metal arc welding (GMAW) (metal inert gas (MIG) welding)
metal transfer processes, in arc welding 53–54
  globular transfer 53
  hot start and cold start transfers 54
  pulsed transfer 54
  short-circuiting transfer 54
  spray transfer 53
microstructure-mechanical property relationship 189–190
MIG *see* gas metal arc welding (GMAW) (metal inert gas (MIG) welding)
misalignment defects 74, **74**, *75*
mobile equipment, application of welding in
  hybrid laser-arc welding 192
  laser beam welding 174
molybdenum 95

N

nanometal matrix composite (MMC) powders 220
Nd:YAG pulsed lasers 170–171, 174–175
new materials processes 218–220
  advancements 219
  nanometal matrix composite (MMC) powders 220
  regulated metal deposition 219
  spray transfer 219
  surface tension transfer 220
nitrogen alloyed stainless steels 93
nitrogen-based steels 14
non-heat-treatable aluminum alloys 154

O

orbital friction welding (OFW) 12, 215
Orowan mechanism 189
overlap defects 65, *65*, **66**
oxy-acetylene gas welding 7–8

P

penetration defects 68, *69*, **70**
phased array ultrasonic testing 85
pilot arc, in arc welding 42
pitting resistance equivalent (PRE) 95
plasma arc welding (PAW) 13, 42, 214
  -based WAAM 201–202
  plasma arc torch 43
plasma formation, in arc welding process 38
plasma non-transferred arc 201
plasma-transferred arc 201
pollution 221
polycrystalline cubic boron nitride (PCBN) tools 11, 157
polymer-based additive manufacturing 200
porosity **64**, 64–65, *65*
powder bed fusion (PBF) 200
precipitate-free zone (PFZ) 130
predictive analytics 207
pressure applied, welding process based on 11
  friction stir welding 11
  friction welding 11–12, *12*
  ultrasonic welding 12, *13*
printing techniques 207
project costs, impact of welding on 227
pulsed gas metal arc welding (P-GMAW) 54
  base current 55
  droplet detachment mechanism 56–57
  peak current 55
  pulse duration 56
  pulse frequency 55–56
  pulsing frequency and duty cycle 54
  thermal behavior of weld 57–61
pulsed laser 25
pulsed transfer, in arc welding 54
pulse-echo testing 85
pulse welding **213**, 215–216

R

radiative heat transfer, in arc welding 53
radiography testing **82**, *83*, 83–84
railways sector, use of welding in 13
real-time process optimization 207
regulated metal deposition welding 219
remote welding operations 221
resistance heating, in arc welding 52–53
  arc voltage 52
  electrode material and diameter 52
  properties of base metal 52–53
  travel speed 52
  welding current 52
resistance welding 6–7, *7*
robotic welding 30–31, 218, 223–224
  advances in 31
  advantages and disadvantages of 31
root hump formation, in laser-arc hybrid welding 188
rotary friction welding (RFW) 11–12, 215

S

safety risks and hazards 225–226
  importance of proper education, training, and equipment 226
  measures to promote occupational health 225
  welding safety 225–226
scope and application of welding 13–14
scratch start method, in arc welding 42
secondary emission 44

selective laser sintering (SLS) 200
seven-axis robots 31
shielded metal arc welding (SMAW) 3–4, *4*, 123–124, 214
shielding gas, in arc welding process 39
shipbuilding sector, use of welding in 13
  laser welding 27
  use of hybrid laser-arc welding in 191–192
short-circuiting transfer, in arc welding 54
silver 92
similar metal joining, using FSW 159–160
skilled workforce, shortage of 224–225
slag inclusion 68–69, **71**, *72*
smart welding **213**, 216–217
smoothed particle hydrodynamics (SPH) 161
solidification cracking
  dissimilar welding of stainless steels 127
  and filler materials 98–99, **100**
    aluminum alloy 104–105, **105**
    carbon and low alloy steel 105–106
    and manganese content in the fusion zone 105–106, *106*
solid-state lasers 26
space technology, use of welding in 14
spatter defects 69, **71**, *72*
splash, in laser welding 27
spray transfer welding 53, 219
stainless steels (SSs) 115–117
  classes 117, *117*
  dissimilar welding of 127–131
    austenitic-austenitic SS 127, *128*
    austenitic-duplex SS 128, *129*
    austenitic-ferritic SS 129, *130*
    duplex-martensitic SS 130
  dissimilar welding of stainless steels (SSs) 127
    austenitic-austenitic SS 127, *128*
    austenitic-duplex SS 128, *129*
    austenitic-ferritic SS 129, *130*
    duplex-martensitic SS 130
  ferritic stainless steel (FSS) and tantalum, joining 92
  joining with other alloys using LBW 173–174
  microstructures 117, **118**
  similar welding of 118
    austenitic SS 118–120, *119*
    duplex SS 123–125, *124*
    ferritic SS 120–123, *122*
    martensitic stainless steel 125–127, *126*
  320 SS and 316L SS, joining 109–112, **110**, **111**
  voltage 116 117
  welding current 116
  welding speed 116
stereolithography 200
stick metal arc welding 3–4, *4*
structure property correlation, welding parameters on 115–117, 131–132
submerged arc welding (SAW) 215
surface tension transfer welding 220

T

tandem GMAW, in additive manufacturing 202
tantalum, joining to ferritic stainless steel (FSS) 92
thermite welding 9
thermo-ionic emission, in arc welding 43
thoriated tungsten 5
3D printing 217, 224
TIG *see* gas tungsten arc welding (GTAW) (tungsten inert gas (TIG) welding)
time-of-flight diffraction 84
titanium 109
touch start method, in arc welding 40–41, *41*
transition zones 101
tungsten (W)-based alloys, and FSW 157
tungsten inert gas (TIG) welding *see* gas tungsten arc welding (GTAW) (tungsten inert gas (TIG) welding)
two-phase flow model 161

U

UFC-MIG technique 146, *146*
ultrasonic peening welding (UPW) 216
ultrasonic testing 84, **84**
  phased array ultrasonic testing 85
  pulse-echo testing 85
  time-of-flight diffraction 84
ultrasonic vibration welding (UVW) 216
ultrasonic welding (USW) 12, *13*, 32–33, 216
  advantages and disadvantages of 33
  working process 33
undercut defects 27, 65, *65*, **66**
underfill defects *65*, 66, **67**
underwater welding 14–15
under welding **213**, 216
Unwin's formula 7

V

virtual reality (VR), for training and simulation 31

W

welding technology, classification of *3*
  based on heat source 3
    electron beam welding 10, *10*
    explosive welding *8*, 8–9
    gas metal arc welding 5–6, *6*

gas tungsten arc welding 4–5, *5*
laser beam welding 9–10
oxy-acetylene gas welding 7–8
resistance welding 6–7, *7*
shielded metal arc welding 3–4, *4*
thermite welding 9
based on pressure applied 11
friction stir welding 11
friction welding 11–12, *12*
ultrasonic welding 12, *13*
wire arc additive manufacturing (WAAM) 217
GMAW-based 202
GTAW-based 201
PAW-based 201–202

Z

zinc-based filler material (Zn-15Al) 93
Zirconiated tungsten 5
ZrB2-7055 joint 145

For Product Safety Concerns and Information please contact our EU representative GPSR@taylorandfrancis.com
Taylor & Francis Verlag GmbH, Kaufingerstraße 24, 80331 München, Germany

www.ingramcontent.com/pod-product-compliance
Lightning Source LLC
LaVergne TN
LVHW010554110826
845149LV00003B/655

* 9 7 8 1 0 3 2 5 6 5 1 3 2 *